Seminary on Geomatics, Civil and Environmental Engineering (2017 BGC)

E3S Web of Conferences Volume 26 (2018)

Gdansk, Poland
22 - 25 June 2017

Editor:

G. Nykiel

ISBN: 978-1-5108-5568-7

Printed from e-media with permission by:

Curran Associates, Inc.
57 Morehouse Lane
Red Hook, NY 12571

Some format issues inherent in the e-media version may also appear in this print version.

Printed by Curran Associates, Inc. (2018)

For additional information, please contact EDP Sciences – Web of Conferences
at the address below.

EDP Sciences – Web of Conferences
17, Avenue du Hoggar
Parc d'Activité de Courtabœuf
BP 112
F-91944 Les Ulis Cedex A
France

Phone: +33 (0) 1 69 18 75 75
Fax: +33 (0) 1 69 28 84 91

contact-edps@webofconferences.org

Additional copies of this publication are available from:

Curran Associates, Inc.
57 Morehouse Lane
Red Hook, NY 12571 USA
Phone: 845-758-0400
Fax: 845-758-2633
Email: curran@proceedings.com
Web: www.proceedings.com

TABLE OF CONTENTS

edp sciences

E3S Web of Conferences

Statement of Peer review

In submitting conference proceedings to E3S Web of Conferences, I certify to the Publisher that I adhere to the **Policy on Publishing Integrity** of the journal in order to safeguard good scientific practice in publishing.

1. All articles have been subjected to peer review administered by the proceedings editors.
2. Reviews have been conducted by expert referees, who have been requested to provide unbiased and constructive comments aimed, whenever possible, at improving the work.
3. Proceedings editors have taken all reasonable steps to ensure the quality of the materials they publish and their decision to accept or reject a paper for publication has been based only on the merits of the work and the relevance to the journal.

Title, date and place of the conference

Seminary on Geomatics, Civil and Environmental Engineering (2017 BGC)
22-25 June 2017
Gdansk University of Technology, Poland

Proceedings editor(s):

Dr Grzegorz Nykiel, Gdansk University of Technology

Date and editor's signature 01/12/2017

Kinetics of nitrogen removal processes in constructed wetlands

Magdalena Gajewska[1,*], *Katarzyna* Skrzypiec[1]

[1]Gdańsk University of Technology, Faculty of Civil and Environmental Engineering, Narutowicza St. 11/12, 80-233, Gdańsk, Poland

Abstract. The aim of this paper is to present a state-of-the-art review of the kinetics of nitrogen removal in constructed wetlands. Biological processes of nitrogen removal from wastewater can be described using equations and kinetic models. Hence, these kinetic models which have been developed and evaluated allow for predicting the removal of nitrogen in treatment wetlands. One of the most important, first order removal model, which is still applied, was analysed and its rate coefficients and factors were compared. This study also demonstrates the validity of Monod and multiple Monod kinetics, commonly seen today. Finally, a computational example of the reaction kinetics of nitrogen removal was also included in the study.

1 Introduction

Several studies have shown the overall effectiveness of constructed wetlands (CWs) in treating municipal wastewater and in removing nitrogen rather effectively. Nitrogen removal is achieved not only by bacteria, but also by plant uptake, adsorption, where ionized ammonia reacts with the media in subsurface horizontal flow (SSHF) constructed wetlands, and volatilization in surface flow (SF) wetlands, where under high pH values the aquatic ammonium phase transforms to gaseous ammonia [5, 6, 9, 10].

Despite the complexity of all biological processes of nitrogen removal in treatment wetland systems, they can be described by relatively uncomplicated equations. Different mathematical models were used to study the process kinetics of the nitrogen removal, mainly first and second order reaction models. Kinetic models developed from different combinations of first–order, Monod kinetics as well as continuous stirred-tank reactor and plug flow patterns can be applied in the design of constructed wetlands for wastewater treatment.

2 Nitrogen removal in constructed wetlands

2.1 Processes and mechanisms

A number of processes transfer nitrogen compounds from one point to another in wetlands without resulting in a molecular transformation. These physical transfer processes include, but are not limited to the following: particulate settling and resuspension, diffusion of dissolved forms, plant translocation, litterfall, ammonia volatilization, and sorption of soluble nitrogen on substrates. In addition to the physical translocation of nitrogen compounds in wetlands, five principal processes transform nitrogen from one form to another: ammonification (mineralization), nitrification, denitrification, assimilation, and decomposition. Due to the low oxygen and organic matter concentration in the root-zone of subsurface flow constructed wetlands, which are limiting factors for nitrification and denitrification processes, a combination of partial nitrification and anaerobic ammonium oxidation has also been suggested as possible pathway for nitrogen removal in constructed wetlands. Because the Anammox process is autotrophic, the complete conversion of ammonium to nitrogen gas can take place without the addition of organic matter. A detailed understanding of these nitrogen transfer and transformation processes is important for understanding wetland treatment systems [1, 10, 12, 17].

2.2 Role of plants in nitrogen removal

There are many effects of vegetation on nitrogen processing and removal in treatment wetlands. Firstly, the plant growth cycle seasonally stores and releases nitrogen, thus providing a "fly-wheel" effect for nitrogen removal time series. Moreover, stable residuals which are created and accreted in the wetlands contain nitrogen as part of their structure, and hence accretion represents a burial process for nitrogen. Another example is submersed litter and stems that provide surfaces on which microbes residue. These include nitrifiers and denitrifiers. The presence of vegetation also influences the supply of oxygen to the water. Emergent vegetation blocks the wind, and shades out algae, presumably lowering reaeration. Floating vegetation may provide a barrier to atmospheric oxygen transfer. Submersed vegetation may provide photosynthetic oxygen supply directly in the water. To some extent, plant oxygen flux supplies protective nitrification in the immediate vicinity of plant roots. Last but not least, carbon content of plant

* Corresponding author: mgaj@gda.edu.pl

litter supplies the energy need for heterotrophic denitrifiers [4, 8].

3 Kinetic models

3.1 First-order kinetics - rate coefficients (*k*)

The speed of pollutant removal is typically represented by a rate coefficient, k:

$$J = k \cdot C \quad (1)$$

where: C = concentration, $g \cdot m^{-3}$, J = removal per unit area, or load removed, $g \cdot m^{-2} \cdot d^{-1}$, k = rate coefficient, $m \cdot d^{-1}$.

This parameter is dependent on factors such as pollutant loading, oxygen transfer, presence of vegetation, etc. Different wetlands will have different k rates, but central tendencies in k can be observed. Most processes in wetlands are dependent on wetland area, so k-rates uses in the wetland literature are often area-based coefficients (k_A). In some intensified wetlands (such as aerated or fill-and-drain wetlands) treatment is dependent on reactor volume, so volume-based rate coefficients (k_V) are sometimes used. K-rates are usually corrected to a temperature standard of 20 °C in the literature [2, 10].

The majority of the models of constructed wetlands are focused on input-output data and the production of either linear regression equations or first order decay laws. The most common form of the first order equations is presented by following equation:

$$C_{out}/C_{in} = \exp(-k_A/q) \quad (2)$$

where q is the hydraulic loading rate in $m \cdot d^{-1}$ and k_A the decomposition constant in $m \cdot d^{-1}$. Another form of the first order equation, which uses the hydraulic residence time (HRT, t) in days, is presented in (3):

$$C_{out}/C_{in} = \exp(-k_v t) \quad (3)$$

The temperature (T) effect in foregoing equations is expressed by the constant k_T (k_A or k_v), which is determined by the use of an Arrhenious equation, as follows:

$$k_T = k_{20} \theta^{(T-20)} \quad (4)$$

where k_{20} ($k_{A,20}$ or $k_{v,20}$) is the value of k_T at 20 °C. The k_{20} and θ constants usually result from a statistical analysis of the data used in model production. Typical values for k_A are between 0.06 and 0.16 $m \cdot d^{-1}$, for k_v between 0.007 and 0.10 $m \cdot d^{-1}$ and the suggested value for θ is 1.05. If $\theta > 1$, reaction rates slow down with colder water temperatures. If $\theta < 1$, reaction rates increase with colder water temperatures [1, 2].

3.2. Background concentrations (C^*)

Internal physical, chemical, and biological treatment cycles within wetlands process, store and release different compounds, this results in non-zero background concentrations for certain compounds. This particularly affects organic matter, nitrogen, phosphorus, and pathogens. Background concentrations may vary on an annual or seasonal basis, depending on the internal cycles involved. These non-zero background concentrations are commonly denoted by the term C^* [2, 10].

3.3 *P-k-C** modelling

Real-world wetlands do not follow the idealized Plug Flow or CSTR (Continuously Stirred Tank Reactor) models. Some short-circuiting and dispersion always occurs in real-world wetlands. This is best represented by a Tank-In Series (TIS) flow model, which is an intermediate case between the Plug Flow and CSTR extremes. In Tank-In-Series (TIS) Flow Model, the number of tanks-in-series (N) is a mathematical representation of the hydraulic performance of a wetland, not the physical number of cells in the system. $N = \infty$ is equivalent to plug flow and $N = 1$ is equivalent to a CSTR [2].

The concepts of non-ideal flow and pollutant weathering (P), first-order kinetics (k) and non-zero background concentrations (C^*) can be combined into a single equation to predict wetland treatment performance. First order plug flow reaction kinetics are commonly used to describe pollutant removal in wetland systems, particularly for BOD_5 and nitrogen removal. The first order plug flow equation incorporating background concentrations is described in (5). It is modified first order k-C^* model in which the concentration varies with residence time t through the constants C_0 the initial concentration, C^* a residual concentration, and k the volumetric rate constant. Because the k-C^* model contains both exponential and linear terms, it cannot be transformed into a simple linear model for statistical calibration.

$$(C - C^*)/(C_i - C^*) = \exp(-kt) \quad (5)$$

where C = effluent concentration, C_i = influent concentration, C^* = background concentration, k = kinetic rate constant. Background concentration is a portion of the C in that is immediately settled down at the unit inlet, and cannot be removed; it is of the order 10–20 $mg \cdot L^{-1}$ [1, 2, 11, 14].

It has been proven that the first order plug-flow kinetic model with background concentrations effectively describes both TKN and NH_4-N removal rates from the SSF and SF wetlands and pond cells with correlation coefficients ranging from $R^2 = 0.52$ to $R^2 = 0.86$ [11].

In order to derive k-values from the available data, the first order P-k-C* model presented by Kadlec and Wallace [10] was used:

$$(C_0 - C^*)/(C_i - C^*) = 1/(1 + k/Pq)^P \quad (6)$$

where C_0 = outlet concentration, $g \cdot m^{-3}$; C_i = inlet concentration, $g \cdot m^{-3}$; C^* = background concentration, $g \cdot m^{-3}$; k = first-order areal rate coefficient, $m \cdot year^{-1}$; P =

apparent number of tanks in series; and q = hydraulic loading rate, m·year^{-1}. The P-k-C* model has been proposed by as being generally the most appropriate kinetic model for predicting outlet concentrations of pollutants displaying first order removal in treatment wetlands. This model incorporates an aspect of the wetland internal hydraulic characteristics in the form of the parameter P, which represents the degree of mixing and approximates the number of completely mixed tanks connected in series [5, 10].

As mentioned before, the most commonly used equation for TN removal in subsurface flow constructed wetlands is the first order equation. According to this approach, the hydrodynamic operation of the wetlands resembles that of plug-flow reactors (PFR), and there is a first order decay of TN with time. Using the TN removal definition [1]:

$$R_{TN} = (C_{in} - C_{out})/C_{in} \quad (7)$$

the right hand side of (7) becomes:

$$1 - R = \exp(-k_v t) \quad (8)$$

3.4 Design features

To estimate the surface area of the SSHF system (A_{HF}) needed to reduce total nitrogen concentration to the target value using empirically determined first-order degradation rate constant (K_{TN}) and background TN concentrations, following equation shall be used:

$$A_{HF} = (Q_{av}/K_{TN}) \ln((C_i - C^*)/(C_0 - C^*)) \quad (9)$$

where A_{HF} = required surface area of bed (m^2), Q_{av} = average water flow rate (m^3·d^{-1}), C_0 = effluent target TN concentration (mg·L^{-1}), C_i = influent TN, C^* = background TN concentration (mg·L^{-1}), K_{TN} = first-order areal rate constant (m·d^{-1}). The rate constant can be set to: K_{TN} = 12 m·year^{-1}. The background concentration for total N can be set to 2 mg·L^{-1} [3, 16].

3.5 Application of artificial neural networks

Artificial neural networks are able to model TN removal in horizontal subsurface flow constructed wetlands. The performance of the networks was found to be reasonably good for wetland design purposes. A design equation for TN removal is proposed as an alternative to the first order model. This equation is based on the ANN (artificial neural networks) response curves, and it reads:

$$R_{TN} = HRT/(K+HRT) \quad (10)$$

with

$$K = (22.8/T)\, 45.5(n/(1 - n))^3 \quad (11)$$

where HRT and K are in days, n is the porosity and $(n/(1-n))^3$ is an expression entering many formulas predicting hydraulic conductivities in porous media. The hyperbolic equation above combines zero and first order kinetics, as this is considered most handy for design of wetlands [1].

3.6 Models based on Monod and multiple Monod kinetics

So far, a few models have been established by combining first-order, Monod and multiple Monod kinetics with continuous stirred-tank reactor (CSTR) flow pattern. Monod kinetics, which contains the parameters of substrate concentration (C_{out}), half saturation constant of limiting substrate (C_{half}) and maximum pollutant removal rates (K_{max}), can be expressed as:

$$dC/dt = -K_{max}(C_{out}/(C_{half} + C_{out})) \quad (12)$$

Combining CSTR flow pattern with Monod kinetics gives (13), which links inlet and outlet pollutant concentrations:

$$(C_{in} - C_{out})/\tau = -K_{max}(C_{out}/(C_{half} + C_{out})) \quad (13)$$

Equation above is expressed in terms of hydraulic retention time τ (units in days). The relation between hydraulic retention time, area (A, m^2), depth (h, m), porosity (ε) of packed media and inlet discharge (Q, m^3·d^{-1}) can be expressed in (14).

$$\tau = Ah\varepsilon/Q \quad (14)$$

Combining volumetric maximum pollutant removal rates (K_{max}, g·m^{-3}·d^{-1}) with h and (ε) results areal maximum pollutant removal rates (K_2, g·m^{-2}·d^{-1}) to correlate inlet and outlet pollutant concentrations.

$$K_2 = q(C_{in} - C_{out})(C_{half} + C_{out})/C_{out} \quad (15)$$

Multiple Monod kinetics assumes that more than one substrate can limit the rate of pollutant degradation. The multiple Monod kinetics can be expressed in the following equation:

$$\frac{dC}{dt} = -K_{max}\left(\frac{C_{out1}}{C_{half1} + C_{out1}}\right)\left(\frac{C_{out2}}{C_{half2} + C_{out2}}\right) \quad (16)$$

C_{out1} and C_{out2} represent the outlet concentrations of two substrates that may limit the biodegradation rate of target pollutant, while C_{half1} and C_{half2} are the half saturate constants for these substrates. Eq. (17) links inlet and outlet pollutant concentration and K_3 value (maximum areal pollutant removal rate K_3, g·m^{-2}·d^{-1}).

$$K_3 = \frac{q(C_{in} - C_{out})(C_{out1} + C_{half1})(C_{out2} + C_{half2})}{C_{out1}C_{out2}} \quad (17)$$

The kinetic coefficients (derived from Monod/multiple Monod kinetics) tend to increase with pollutant loading, indicating that the coefficients may vary based on different factors, such as influent pollutant concentration, hydraulic loading, and water depth [13, 15, 16].

4 Calculation example

The reaction kinetics calculations were performed for the SF (Surface Flow) wetland system.
• Influent flow rate (Q) = 180 $m^3 \cdot d^{-1}$ = 65700 $m^3 \cdot y^{-1}$,
• Influent BOD (C_i) = 300 $mg \cdot L^{-1}$,
• SF area (A) = 1 ha = 10000 m^2.

What is the effluent concentration (C) at 20 °C? Assume k_A = 36 $m \cdot y^{-1}$ and P = 3.

$$\frac{C - C^*}{C_i - C^*} = \frac{1}{(1 + k/Pq)^P}$$

$$q = \frac{Q}{A} = \frac{65700 m^3 \cdot y^{-1}}{10000 m^2} = 6.57 m \cdot y^{-1}$$

Background concentrations in SF wetlands for BOD can be expressed as an empirical relationship based on curve fitting to existing data sets:

$$C^* = 0.6 + 0.065C_i = 0.6 + 0.065 \cdot 300 = 20.10 \text{ mg} \cdot L^{-1}$$

Effluent concentration at 20 °C is:

$$C = \frac{C_i - C^*}{(1 + k/Pq)^P} + C^* = \frac{300 - 20.10}{(1 + 36/3 \cdot 6.57)^3} + 20.10 = 32.50 mg \cdot L^{-1}$$

What is the effluent concentration (C) at 10 °C? Assume k_A = 37 $m \cdot y^{-1}$ at 20 °C and θ = 0.985.

$$k_{T=10^oC} = k_{20}\theta^{(T-20)} = 37 \cdot 0.985^{(10-20)} = 43 m \cdot y^{-1}$$

Effluent concentration at 10 °C is:

$$C = \frac{C_i - C^*}{(1 + k/Pq)^P} + C^* = \frac{300 - 20.10}{(1 + 43/3 \cdot 6.57)^3} + 20.10 = 27.78 mg \cdot L^{-1}$$

With a slight change in the k coefficient at lower temperature (10 °C), a greater reduction in BOD concentration occurs than at 20 °C. This confirms that when $\theta < 1$and at lower temperatures, the reaction rates increase.

5 Summary

First-order, area-based nitrogen loss models provide a suitable method for design of wetland treatment systems in most circumstances. These have the advantage of correctly describing internal phenomena in flow through wetlands, as well as describing batch wetland operation. Studies on side-by-side wetlands confirm the effects of the principal variables of inlet concentrations and hydraulic loading rates (or the equivalent detention times). The parent mass balance equation for water movement may be adjusted to fit extreme environmental conditions of precipitation or evapotranspiration. The rate equations account for return fluxes from the wetland biomass, and thus can fit the entire range of hydraulic loadings. In parameter estimation, the sequential nature of the nitrogen transformations cannot be ignored.

References

1. C.S. Akratos, J.N.E. Papaspyros, V.A. Tsihrintzis, Bioresour. Technol. **100**, 586 (2009)
2. C.A. Arias, T.R. Headley, P. Carvalho, *Constructed Wetlands For Water Pollution Control – PhD training course materials*, 18–24 (Aarhus University, Denmark, 2017)
3. H. Brix, N.H. Johansen, *Nutrient Cycling and Retention in Natural and Constructed Wetlands*, 155 (Publishers, Leiden, The Netherlands, 1999)
4. J. Fan, W. Wang, B. Zhang, Y. Guo, H.H. Ngo, W. Guo, J. Zhang, H. Wu, Bioresour. Technol. **143**, 461 (2013)
5. M. Gajewska, H. Obarska-Pempkowiak, Environ. Prot. Eng. **37**, 27 (2011)
6. M. Gajewska, K. Ambroch, Pol. J. Environ. Stud. **21**, 65 (2012)
7. T.R. Headley, C.C. Tanner, Crit. Rev. Environ. Sci. Technol. **42**, 2261 (2012)
8. R.H. Kadlec, *Natural and Constructed wetlands: Nutrients, Metals and Management*, 233 (Backhuys Publishers, Leiden, The Netherlands 2005)
9. R. Kadlec, R. Knight, *Treatment Wetlands* (CRC Press, 1996)
10. R.H. Kadlec, S. Wallace, *Treatment Wetlands, Second Edition* (CRC Press, Boca Raton 2009)
11. C.B. Kinsley, A.M. Crolla, N. Kuyucak, M. Zimmer, A. Lafleche, Water Sci Technol. **56**, 151 (2007)
12. D. Paredes, P. Kuschk, F. Stange, R.A. Müller, H. Köser, Water Sci Technol. **56**, 145 (2007)
13. T. Saeed, G. Sun, Bioresour. Technol. **102**, 1205 (2011)
14. O.R. Stein, B.W. Towler, P.B. Hook, J.A. Biederman, Water Sci Technol. **56**, 93 (2007)
15. G. Sun, T. Saeed, Process Biochem. **44**, 717 (2009)
16. J. Vymazal, H. Brix, P.F. Cooper, R. Haberl, R. Perfler, J. Laber, *Removal mechanisms and types of constructed wetlands* (Backhuys Publisher, 1998)
17. Y.Q. Zhao, G. Sun, S.J. Allen, Water Res. **38**, 2907 (2004)

Analysis of efficiency of phosphates sorption by different granulation of selected reactive material

Magda Kasprzyk[1,*], *Jarosław* Węgler[1], and *Magdalena* Gajewska[1]

[1]Faculty of Civil and Environmental Engineering, Gdansk University of Technology, Poland

Abstract. In the light of the need to find an effective way to remove phosphorus from wastewater, studies on the suitability of sorption materials in this process should be conducted. The aim of the study was to examine the potential benefits of using selected adsorbents to reduce orthophosphates from the model solution under steady conditions. The study was conducted on a laboratory scale using synthetic wastewater with concentration of P-PO_4 in the range of 5-100 mg/dm3. Experiment has shown that fine-grained material M1 (0-2 mm) is highly effective at removal of phosphorus compounds at the level of 97.8% at the highest concentration of P-PO_4. The sorption capacity achieved during the investigation was 9.6 mg/g, while the maximum sorption capacity from the Langmuir model could reach up to 256 mg/g. Material M2 (2-8 mm) did not show satisfactory sorption capacity (maximum calculated sorption capacity: 0.36 mg/g) and the effectiveness of phosphate reduction did not exceeded 6% at the lowest concentration of P-PO_4.

1 Introduction

Phosphorus, as a nutrient element, together with the nitrogen compounds enable plants growth. However, these processes require a small amount of nutrients. Their excess is the cause of eutrophication and algae bloom. Phosphorus compounds originate mainly from wastewater discharged into the lakes and other water bodies [1].

The removal of nutrients from wastewater is one of the basic problems of wastewater treatment plants. Most often their level is reduced in biological processes (using phosphorus as a source of energy for microorganisms) and chemical processes (reducing phosphate by salts of ferric or aluminum) of wastewater treatment [2].

Wastewater containing significant amounts of phosphorus compounds may originate both from large wastewater treatment plants and from smaller rural or domestic wastewater treatment plants. Most often the small wastewater treatment plants may discharge to the receivers wastewater purified insufficiently. The solution may be the use of filtration materials in domestic wastewater treatment plants and constructed wetlands [3, 4].

In recent years, the subject of research in terms of effectiveness are sorption materials (adsorbents) capable to bind significant amounts of phosphorus compounds. The source of their origin can be natural and anthropogenic. They may consist of silicon, calcium, aluminum, sulfur, iron in various weight ratio [5-12].

Unfortunately, the research methodology for evaluating the effectiveness of adsorbents in process of wastewater treatment is still under discussion. The lack of standardization the guidelines is a cause of difficulties in comparing different materials and their application in practice. In accessible research papers can be found methods differ in the concentration of the model solution, the volume, the amount of added sorption material, way and mixing time of the sample and the time of contact adsorbent with the model solution. Therefore, the scope of the differences in research methodology is significant [1, 13].

The aim of this study was an evaluation of selected sorption material in two fractions to compare the influence of grain size on phosphorus removal efficiency from wastewater in steady conditions. Another purpose was the determination of parameters of the Langmuir and Freundlich adsorption isotherms and maximum sorption capacity.

2 Materials and methods

2.1 Materials

Fig. 1. Material M1 – 0-2 mm (on the left), material M2 – 2-8 mm (on the right).

* Corresponding author: magkaspr@pg.edu.pl

Material M1 (Fig.1) is a fine-grained (0-2 mm) product of thermal treatment of carbonate-siliceous rock with high content of calcium carbonate $CaCO_3$ in temperature 700°C. This fraction if by-product of the M2 material and due to very fine-grain consisting mostly a calcium oxide CaO (above 80 %).

Material M2 is produced in process of decarbonization of natural carbonate-siliceous rock (opoka) (Fig.1). Because of high content of elements like calcium, silicon and aluminum, material M2 presents abilities to bind phosphorus compounds from wastewater. Material M2 is also known under the name ROCKFOS®. The only Polish producer of Rockfos® is Ceramika KUFEL from Suchynia.

Calcium oxide CaO and silicon oxide SiO_2 present in the Rockfos® chemical composition in the highest concentration (Table 1).

Table 1. Chemical composition of Rockfos® [20].

Chemical composition	Concentration
CaO	43.3%
SiO_2	36.0%
Al_2O_3	5.9%
Na_2O	2.9%
Fe	1.3%
TiO_3	0.96%
MgO	0.94%
S	0.65%
∑ K_2O, P, Cl, MnO	~2%

Porosity of Rockfos® is over 50 % and size of particles is around 2-8 mm. Disadvantage of this material may be particularly high level of pH (11-12), which results in alkaline solution (Table 2).

Table 2. Physical characteristic of Rockfos® [20].

Physical properties	Value
Average particle size (mm)	2–8
Porosity (%)	~50
pH	11– 12

2.2 Experiment design

To carry out the laboratory investigation, portions of 10 g of each selected materials was prepared. In order to determine the adsorption isotherms of these materials model solution was used. The model solution was prepared from distilled water and a solution of potassium dihydrogen phosphate KH_2PO_4 (1 cm^3 - 10 $mgPO_4/dm^3$). In each of the 5 batch reactors, there was a model solution in a volume of 1.0 dm^3 with a phosphate phosphorus concentration of 5-100 mg/dm^3 (Table 3).

To each of the 5 batch reactors, 10 g of selected material was added. Mixing time lasted 5 minutes and was provided by magnetic stirrer (1000 rpm). Samples were collected after 1 hour of sedimentation. Three series of repetitions were made for each material. All tests were performed at room temperature of approximately 20 °C.

Table 3. Concentrations of P-PO_4 in input samples ("0").

Batch reactor		1	2	3	4	5
Assumed initial concentration [mg/dm³]		5.0	10.0	20.0	50.0	100.0
Average real concentration [mg/dm³]	M1	5.11	9.94	19.6	48.1	97.3
	M2	5.11	10.05	19.6	47.9	96.0

2.3 Physical and chemical analysis

Before the designation of the concentration of P-PO_4 of synthetic wastewater, the samples were subjected to filtration. The concentration of the phosphorus phosphate was marked by using the cuvette tests HACH Lange and HACH Lange DR 3900 laboratory VIS spectrophotometer with RFID.

On the basis of the obtained results, the series of analyses were conducted in order to verify the suitability of selected materials for removing phosphorus compounds by adsorption. The effectiveness of binding P-PO_4 by selected materials was calculated by equation (1) [13]:

$$E = (C_0 - C) / C_0 \cdot 100\% \qquad (1)$$

where: E – effectiveness of phosphorus removal [%], C_0 – initial concentration of P-PO_4 in model solution [mg/dm^3]; C – final concentration of P-PO_4 in model solution [mg/dm^3].

Quantity of adsorbed amount of phosphorus compounds from model solution was designated according to formula (2) [12, 14]:

$$q = (C_0 - C) \cdot V / m \qquad (2)$$

where: q – quantity of adsorbed P-PO_4 from model solution [mg/g], V – volume of model solution [dm^3], C_0 – initial concentration of P-PO_4 in model solution [mg/dm^3]; C – final concentration of P-PO_4 in model solution [mg/dm^3], m – mass of sorption material [g].

2.4 Adsorption models

Sorption capacity was analyzed on the basis of the relation between the quantity of adsorbed phosphorus by sorption material q_e [mg/g] and equilibrium concentration C_e [mg/dm^3]. This relation is described by mathematical equations of adsorption isotherms. Parameters of the adsorption isotherms of selected materials were determined based on model given by [13,15-18]:

a) Langmuir

$$q_e = (K_L \cdot C_e) / (1 + a_L \cdot C_e) \qquad (3)$$

where: q_e – sorption capacity [mg/g]; a_L [dm^3/mg], K_L [dm^3/g] – constant parameters in Langmuir model of adsorption; C_e – concentration of P-PO_4 at equilibrium [mg/dm³], $K_L/a_L = q_{max}$ – maximum sorption capacity [mg/g]. Equation (3) in linear form becomes:

$$1/q_e = 1/K_L \cdot 1/C_e + a_L/K_L \quad (4)$$

b) Freundlich

$$q_e = a_F \cdot C_e^{bF} \quad (5)$$

where: q_e – sorption capacity [mg/g]; C_e – concentration of P-PO_4 at equilibrium [mg/dm³]; a_F [dm³/g], b_F [dm³/mg] – constant parameters in Freundlich model of adsorption. Equation (5) in linear form becomes:

$$\log q_e = \log a_F + b_F \cdot \log C_e \quad (6)$$

Graphical interpretation of linear equations (4) and (6) are shown in figures 2-5, where adsorption isotherms are presented in form (7):

$$y = a \cdot x + b \quad (7)$$

where:

for Langmuir isotherm:

$y = 1/q_e$ $x = 1/C_e$ $a = 1/K_L b = a_L/K_L$

for Freundlich isotherm:

$y = \log q_e$ $x = \log C_e$ $a = b_F$ $b = \log a_F$

3 Results and discussion

The basic parameters (q, E) to evaluate the efficiency of phosphorus removal from synthetic wastewater of two selected materials are presented in Table 4.

Table 4. Sorption capacity and effectiveness of phosphorus removal.

M1			
C_0 [mg/dm³]	C [mg/dm³]	q [mg/g]	E [%]
5.11	0.27	0.49	94.8
9.94	0.53	0.94	94.7
19.6	0.9	1.9	95.4
48.1	1.5	4.7	97.0
97.3	2.1	9.6	97.8
M2			
C_0 [mg/dm³]	C [mg/dm³]	q [mg/g]	E [%]
5.11	4.90	0.03	6.0
10.05	9.52	0.05	5.3
19.6	18.6	0.1	5.4
47.9	46.5	0.1	2.9
96.0	93.0	0.3	3.1

For material M1 efficiency of phosphorus reduction and sorption capacity increased with the increase of initial concentration of P-PO_4. At the highest concentration of P-PO_4 sorption capacity reached 9.6 mg/g, and the level of reduction of phosphorus compounds was nearly 98 %. Similar results also for fine-grain particles (1-2 mm) were occurred for carbonate-siliceous rock (opoka) heated in temperature 900 °C. Efficiency of phosphorus removal reached to 97 % [2]. Those study was proceeded with flow conditions in models of filters with vertical flow.

Material M2 showed much lower efficiency of P-PO_4 removal, which decreased along with increasing initial concentration. For the lowest concentration (5.11 mg/dm³), the effectiveness of binding phosphates by sorption material was only 6 %, while maximum sorption capacity at the highest concentration was equal to 0.3 mg/g.

Graphical analysis of the results of laboratory investigation showed good matching for both Langmuir and Freundlich isotherms of adsorption to describe the sorption process for both sorption materials (Fig. 3-6). The coefficient of determination for Langmuir isotherm was equal to 0.99, and for Freundlich isotherm was about 0.97 (Table 5). The linear correlation coefficient was obtain on the level of $r = 0.996$ for Langmuir isotherm and $r = 0.985$ for Freundlich isotherm.

Maximum sorption capacity of analyzed materials was obtained from parameters of Langmuir isotherm and equal 294.12 mg/g for material M1, and 0.36 mg/g for material M2. Sorption capacity of both materials is incomparable (Table 5).

Table 5. Parameters of the Langmuir and Freundlich adsorption isotherms of materials M1 and M2.

Langmuir isotherm				
	K_L [dm³/g]	a_L [dm³/mg]	q_{max} [mg/g]	R^2
M1	2.1598	0.0084	256.41	0.991
M2	0.0070	0.0191	0.36	0.990
Freundlich isotherm				
	a_F [dm³/g]	b_F [dm³/mg]	R^2	
M1	2.8438	1.3358	0.968	
M2	0.0104	0.7276	0.972	

Comparing the results of current laboratory investigation, material M1 has shown one of the higher sorption capacity. Similar results were obtained in the studies presented in [12], where the sorption material was composed mostly of calcium oxide (70 %), reached maximum sorption capacity over 285 mg/g. Material composed of calcium sulfate, magnesium and iron compounds obtained q_{max} close to 160 mg/g. Methodology of the research was different from that presented in this paper, as contact time, which lasted 24 hours. According to [6] the highest sorption capacity was achieved for Polonite® (94.32 mg/g), but in this case, contact time was only 15 minutes.

Also very high sorption capacity was achieved in the experiment described in [19] at the level of 182 mg/g. The contact time was also 1 hour. Examined materials were different types of opoka heated/treated in temperature 900 °C. The highest sorption capacity was received for opoka with the highest calcium content (420 g/kg). In process of high temperature treatment the conversion of calcium carbonate to calcium oxide occurred and released of carbon dioxide. Increased concentration of calcium oxide was causing the increase of sorption capacity. Also, pH of opoka was rising from 6.8 to over 12. In article [8] various types of heated opoka were analyzed. Obtained results of sorption capacity were around 12-25 mg/g with short contact time which lasts – 15 minutes. Again the best results were received for opoka with the highest content of CaO (520.1 g/kg).

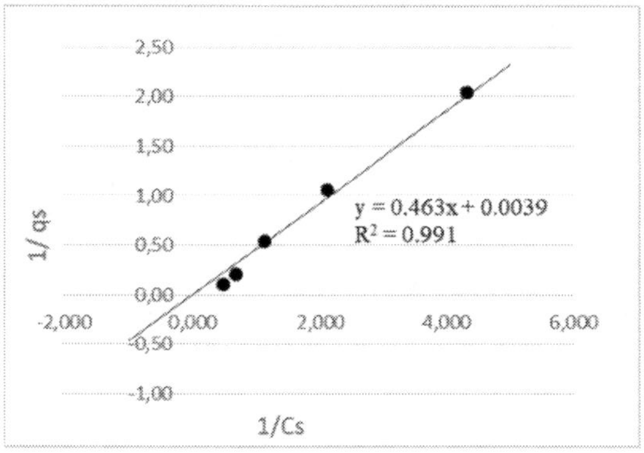

Fig.2. Langmuir isotherm of material M1.

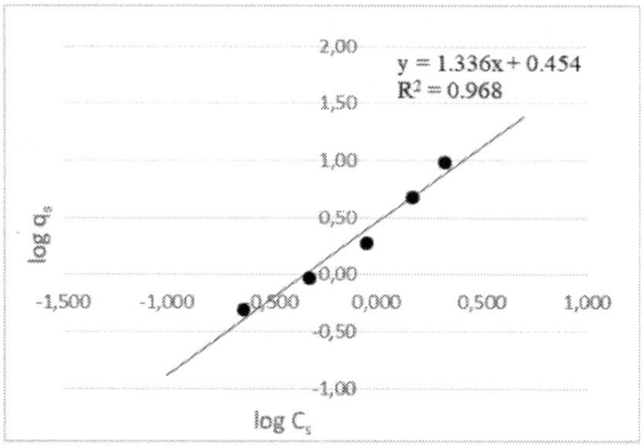

Fig.4. Freundlich isotherm of material M1.

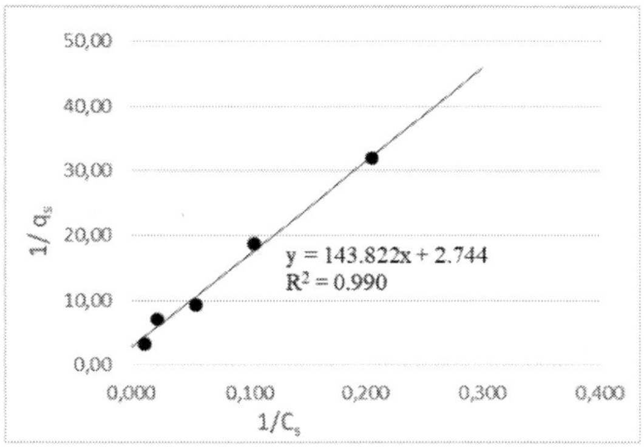

Fig.3. Langmuir isotherm of material M2.

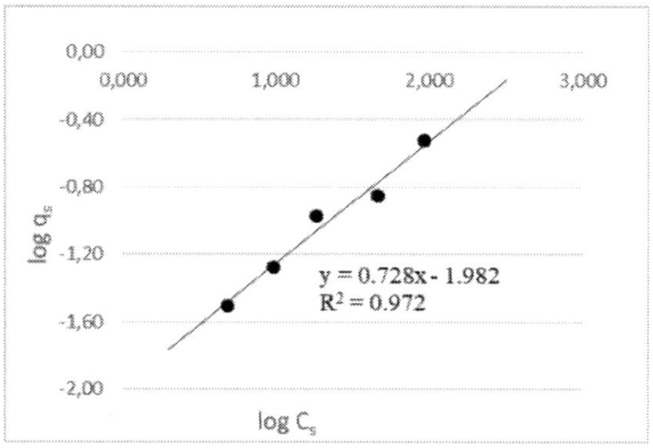

Fig.5. Freundlich isotherm of material M2.

4 Conclusions

Based on obtained results and their evaluation, following statements could be concluded:

1. The achieved results confirmed the influence of both grain size of the sorption material as well as the conditions (flow or stable) for removal efficiency.
2. In stable condition, the material M2 was ineffective (maximum sorption capacity was only 0.36 mg/g) and can be considered as unsuitable for the removal of phosphorus from wastewater in such conditions.
3. Material M1 has shown substantial ability to remove phosphorus compounds (256.4 mg/g) and also was highly effective in adsorption process – 97.8 %. Preliminary analysis allows to apply this material in further studies. Also because of the high content of calcium oxide, those studies should consider the influence of material over pH changes.
4. Investigation has confirmed the influence of content of calcium oxide and size of the particles over achieved sorption capacity level. With the higher content of CaO and smaller size of the particles, sorption capacity is increasing.
5. Used mathematical models of adsorption (Langmuir and Freundlich) has shown well matching of isotherms for graphical analysis of received results ($R^2 = 0.97$-0.99).
6. There is a need to test the applicability of M2 material in flow conditions for phosphorus removal.
7. Also there is a necessity to standardize methodology to determine the sorption capacity of materials used in various experiments. They should describe the way of evaluation suitability to remove phosphorous compounds from wastewater.

The research is carried out within the subtask 2.3 of the project entitled "Integrated technology for improved energy balance and reduced greenhouse gas emissions at municipal wastewater treatment plants" with the acronym "BARITECH" co-funded by the Norwegian funds, under the Polish-Norwegian Cooperation Research carried out by the National Centre for Research and Development [197025/37/2013].

Special thanks to Polish producer of Rockfos®, Ceramika KUFEL from Suchynia for the free of charge sharing of research material.

References

1. L. Johansson Westholm, Water Research **40 (1)**, 23 (2006)
2. K. Jóźwiakowski, M. Gajewska, A. Pytka, M. Marzec, M. Gizińska-Górna, A. Jucherski, A. Walczowski, M. Nastawny, A. Kamińska, S. Baran, Ecological Engineering **98**, 290 (2017)
3. A. Jucherski, M. Nastawny, A. Walczowski, K. Jóźwiakowski, M. Gajewska, Ochrona Środowiska **39 (1)**, 33 (2017)
4. H. Brix, C.A. Arias, M Del Bubba, Water Science and Technology 44 (11-12), 47 (2001)
5. G. Renman, A. Renman, *WASCON 2012 Conference proceedings* (2012)
6. A. Karczmarczyk, A. Bus, Annals Of Warsaw University Of Life Sciences – SGGW. Land Reclamation **46 (1)**, 57 (2014)
7. A. Bus, A. Karczmarczyk, A. Baryła, Inzynieria Ekologiczna **39**, 33 (2014)
8. A. Bus, A. Karczmarczyk, Infrastruktura i Ekologia Terenów Wiejskich **II/1**, 227 (2014)
9. C. Vohla., M. Kõiv, H.J. Bavor, F. Chazarenc, Ű. Mander, Ecological Engineering **37**, 70 (2011)
10. A. Karczmarczyk, A. Bus, A. Baryła, Ecological Engineering **86**, 69 (2016)
11. A. Renman, On*-site wastewaters treatment – Polonite and other filter materials for removal of metals, nitrogen and phosphorus* (TRITA-LWR PhD Thesis 1043, KTH Royal Institute of Technology, 2008)
12. M. Nastawny, A. Jucherski, A. Walczowski, K. Jóźwiakowski, A. Pytka, M. Gizińska-Górna, M. Marzec, M. Gajewska, A. Marczuk, J. Zarajczyk, Przemysł Chemiczny **94 (10)**, 1001 (2015)
13. A. Bus, A. Karczmarczyk, Fresenius Environmental Bulletin **24 (9)**, 2736 (2015)
14. X. Liu, L. Zhang, Powder Technology **277**, 112 (2015)
15. V. Cucarella, G. Renman, J. Environ. Qual. **38**, 381 (2009)
16. M. Del Bubba, C.A. Arias, H. Brix, Water Research **37 (14)**, 3390 (2003)
17. Z. Xu, Jian-Guo Cai, Bing-Cai Pan, Journal of Zhejiang University-SCIENCE A, Applied Physics & Engineering **14 (3)**, 155 (2013)
18. G.Limousin, J.P. Gaudet, L. Charlet, S. Szenknect, V. Barthea, M. Krimissa, Applied Geochemistry **22**, 249 (2007)
19. V. Cucarella, T. Zaleski, R. Mazurek: Annals of Warsaw University of Life Sciences – SGGW. Land Reclamation **38**, 11 (2007)
20. Product Data Sheet of Rockfos®

An evaluation of processing InSAR Sentinel-1A/B data for correlation of mining subsidence with mining induced tremors in the Upper Silesian Coal Basin (Poland)

Artur Krawczyk[1,*], *Radosław* Grzybek[1]

[1]AGH University of Science and Technology, Faculty of Mining Surveying & Environmental Engineering; Department of Mining Areas Protection, Geoinformatics and Mining Surveying, al. A. Mickiewicza 30, 30-059 Krakow, Poland

Abstract. The Satellite Radar Interferometry is one of the common methods that allow to measure the land subsidence caused by the underground black coal excavation. The interferometry images processed from the repeat-pass Synthetic Aperture Radar (SAR) systems give the spatial image of the terrain subjected to the surface subsidence over mining areas. Until now, the InSAR methods using data from the SAR Systems like ERS-1/ERS-2 and Envisat-1 were limited to a repeat-pass cycle of 35-day only. Recently, the ESA launched Sentinel-1A and 1B, and together they can provide the InSAR coverage in a 6-day repeat cycle. The studied area was the Upper Silesian Coal Basin in Poland, where the underground coal mining causes continuous subsidence of terrain surface and mining tremors (mine-induced seismicity). The main problem was with overlapping the subsidence caused by the mining exploitation with the epicentre tremors. Based on the Sentinel SAR images, research was done in regard to the correlation between the short term ground subsidence range border and the mine-induced seismicity epicentres localisation.

1 Introduction

The Upper Silesian Coal Basin (USCB) is located in the south of Poland. It is known as the biggest active area of black coal mining in Europe. As a result of underground (mostly longwall) mining, the surface is susceptible to continuous deformations creating a subsidence area. The subsidence is a continuous process that occurs because of and alongside an excavation process. Additionally the Upper Silesian Coal Basin is known for its complicated geological conditions, thus making it one of the most seismically active mining areas in the world. Seismic investigations performed between 1974 and 2005 registered over 55900 mining related tremors of energy $E \geq 105$ J (local magnitude $ML \geq 1.5$) [1]. On account of a constant decrease in mining activities since 1995, a constant reduction in the number of tremors was reported. However, since the excavation process currently covers deeper parts of coal deposits. tremors that occur are of higher energies. The occurrence of some of the tremors not only causes a lot of expensive damage to surface constructions and buildings but also affects underground mining infrastructure (excavation, shafts, drafts, etc.).

The mining tremor is a short-lasting, hard to analyse episode. Its epicentre almost always originates within the subsidence area. On account of that, a hypothesis was formed stating that high energy tremors may result in abrupt land subsidence caused by sudden energy releases within rock mass body. This hypothesis was proven with various land surveying works done in the area of an underground cooper excavation (LGOM - Lower Silesia region in Poland) [2]. In the mentioned area, typical gradient of subsidence was 3 to 4 mm per day [3]. However, in the days following strong tremors, it reached 119 mm per day. Daily measurements performed on survey lines have proven to be extremely expensive. Typically, this kind of survey is performed once a year, however, in some cases it is necessary to conduct them on some lines once every 6 months. It is easy to notice that this kind of data does not provide continuous monitoring of small elevation changes of an area. On account of that, in between 2002 and 2005 a scientific investigation of possible use of the SAR interferometry of LGOM area was performed [4]. The aim was to perform the monitoring of short term increase in height change of the subsidence area. The results of these studies showed that, for 2002-2005 period, available satellite data imaging is not frequent enough to describe effects of tremors that occur often. At that time, data from the Envistat-1 and ERS 2 satellites had minimum time interval of available data at 35 days. A large number of tremors (up to four) observed every month in a single subsidence area prevented from creating a proper way of analysing and monitoring subsidence area that originated from mining tremors. In 2011, with the use of the PSInSAR method, scientific research was conducted on the correlation between ground subsidence and mining induced seismicity within the USCB [5]. Based on long term observation, the correlations was proved.

Mine tremors are divided into two types: 1) the first type of tremor is directly connected with mining

* Corresponding author: artkraw@agh.edu.pl

 https://doi.org/10.1051/e3sconf/20182600003

operations, 2) the second one, which is the seismicity induced by mining, is strongly affected by local geology and tectonics, and interaction between mining, lithostatic and residual tectonic stress on regional level and scale. They usually occur at some bigger distance from mining face and are generally connected to the geological discontinuities [6]. Mining exploitation affects on terrain surface by continuous deformation caused by the subsidence area over exploitation panel.

The aim of the study presented in this article is to evaluate the possibility of recording the short term subsidence caused by mining activities and tremors based on the SAR interferometry. All data should be correlated in time and space. All analyses will be performed with open access data.

2 Area of research and source data characterisation

The studied area for this article is the Upper Silesia Coal Basin (USCB), covering about 5,400 square kilometres. Significant part of the area lies within the borders of Poland (4,500 square kilometres), however, it partly reaches beyond them, into the northeastern territory of Czech Republic. The black coal excavation in this region started about 200 years ago and as at 31.12.2013 the resources were assessed at 51 million tons. The excavation is currently done by 21 underground mines. The estimated terrain subsidence caused directly by mining activities is 800 square kilometres.

The first data source used for this study was obtained from the Synthetic Aperture Radar (SAR-C) placed on the Sentinel-1A and 1B satellites. The Sentinel missions are part of the EC (European Commission) Earth observation programme - Copernicus. The European Space Agency (ESA) is in charge of the technical development and maintenance of the mentioned satellites. A repeat interval (also called a revisit time) of the Sentinel satellites is 12 days, but since 2 satellites occupy opposite sides of their orbit (angular distance of 180 degrees), data of the same area are available in 6 days intervals [7]. On account of the Copernicus agenda, all SAFE (Standard Archive Format for Europe) data are available for the public free of charge.

The second data set used for this study is information on mining tremors occurrence. They were obtained from the stations of the Upper Silesian Seismological Network operated by the Central Mining Institute. Data from this source cover the area of about 2000 square kilometres [8]. The seismological network consists of 15 stations. Basing on various data sets gathered by these stations, basic seismological parameters are calculated. An available data bank provides information on the time of dynamic even (tremor) occurrence, its energy, the magnitude, the name of the mine involved and the epicentre coordinates. It is important to mention that energy of tremors is given in M (magnitude) instead of J (Joule – from International System of Units).

Maps of mining areas are the third type of data necessary for this project. These data are available via the MIDAS system provided by the Polish Geological Institute – the National Research Institute (PGI – NRI). Maps of mining concession areas can be downloaded via WMS (Web Map Service). These data allow to connect the subsidence areas to certain mine. Map from Fig. 1 shows the studied area.

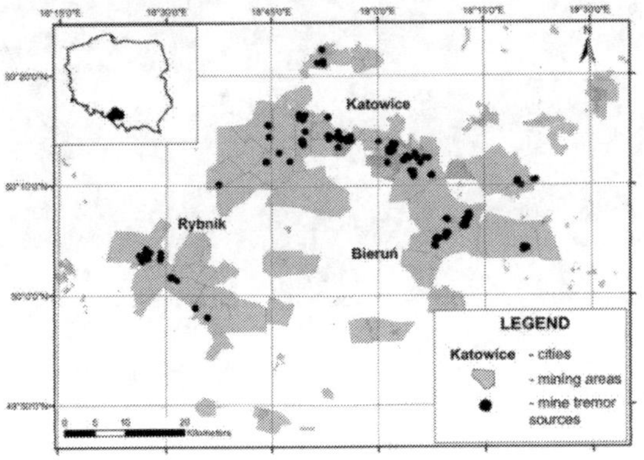

Fig. 1. Studied area including mining areas and epicentres of mining tremors.

3 Measurement method

For the purpose of this research, SAR (Synthetic Aperture Radar) type data acquired by Sentinel-1A/B (S1A/S1B) was downloaded from the Sentinels Scientific Data Hub. The European Space Agency (ESA) also provides an open source software, called SNAP, for processing radar images. This software was used for transforming radar into interferometric images. The Digital Elevation Model, used in subtracting topographic phase from the interferogram, was obtained from the SRTM digital elevation data, produced by NASA, with 3 arc second resolution. The size of interferogram pixel is 11.2 metres.

The SAR interferometry is a well-known technique that provides accurate measurements of terrain elevation changes based on the repeated SAR acquisitions [9,10]. The technique, as well as its applications and limitations, was described well in literature [11]. The terrain subsidence of the USCB area has been already studied by various teams [12,13,14,15].

The SAR interferograms were transformed into the Polish coordinates system - PUWG 1992. Data on mining tremors, provided in the *.xls format, were geocoded into shape layers and imported into the ArcGIS software. Since these data were provided in the geographic coordinates system, it had to be transformed into the PUWG 1992 system. This supplied the users with sufficient information for further analysis. Fig. 2 shows an example of a S1A/S1B SAR interferogram of the subsidence area within a 6-day cycle. It also shows an epicentre of the tremor, a centre of subsidence area and a distance between those two points. Based on this image, it can be estimated whether the tremor epicentre falls within the area of land subsidence or not. On account of degradation of coherence, which was due to environmental changes, not all subsidence areas were detected [7]. Also, in some cases, not all area parts were detected [8], thus not in all cases the distance between the mentioned centres was calculated.

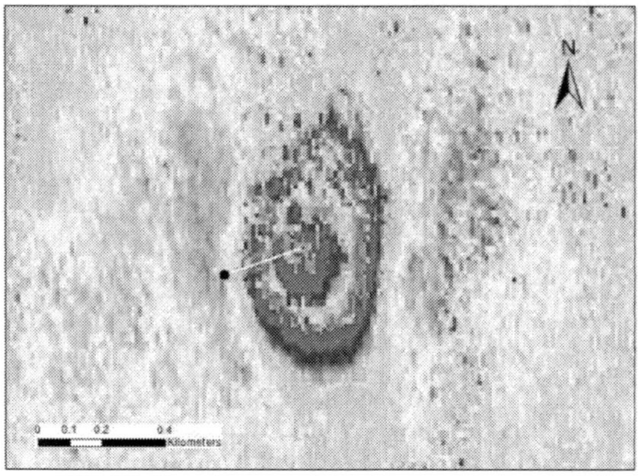

Fig. 2. A SAR interferogram of subsidence area including the distance of 260 m between epicentre of the tremor (occuring at 2016-11-16) and centre of 6 days subsidence area.

The other issue is estimating the value of subsidence trough registered on interferograms in the form of fringes. The SAR image is not registered vertically, but in the angular range of about 34 degrees. It requires recalculating half of wavelength 2.78 cm into a single sequence of fringes and then reducing it due to the incidence angle of radar signal. This allows to estimate the maximum value of subsidence in the area showed in Fig. 2 as 23 mm circa. The problem of estimating the values of subsidence registered on interferograms is a separate research issue and it was not analysed in this article.

4 Conducted experiments and results discussion

The time frame of this study was established between 01.09.2015 and 10.02.2017 and included 528 days. During that time, 303 radar images were acquired by Sentinel-1A and Sentinel-1B. The Upper Silesian Seismological Network registered 318 tremors with the magnitude higher than 2.5 M. Unfortunately, from 01.09.2015 to 27.09.2016, only the data from Sentinel-1A were available, limiting the interferograms to a 12-day repeat cycle.

Because of the unfavourable land cover (woodlands, agriculture etc.) and the fact that small energy (up to 2 M) tremors usually do not cause significant damage in mining infrastructure, only 41 areas were selected for further analysis. For selected areas, the 40 SAR scenes were chosen and processed into interferograms. After analysing 41 tremors, in case of 22 of them, it was possible to link them with the subsidence areas on interferograms. In case of the other 19, it was impossible to perform the analysis because of the coherence loss, most likely caused by atmospheric conditions. Furthermore, within the 22 correctly located subsidence areas, only 13 were coherent enough to allow estimation of its centre and the measurement of the distance from the centre to the tremor epicentre. For the purpose of this article, we defined the geometrical method for selecting tremors by the border of the subsidence area. The second parameter is the distance between the epicentre of tremor and the centre of the subsidence area.

The border for a 6-day subsidence trough can be easy detected from the InSAR image. This area includes the part of rock mass that loses support and sags to fill the void beneath the longwall black coal exploitation system. The mechanism progresses from the coal seam towards the surface and affects the surface in a larger area than the area of the extracted panel of coal that is undergoing the settlement. The subsidence is typically lesser than the thickness of coal extracted underground and the resulting border of subsidence trough is the maximal extend of the rock mass deformation caused by the longwall exploitation system.

Based on the border of subsidence trough, three points and two distances were defined. Both of them are presented on Fig 3. There are: 1) cs - the centre of subsidence trough, 2) sb – the border of the subsidence 3) e - the tremor of the epicentre. The line connects the epicentre with the subsidence. The distance will be measured along this line.

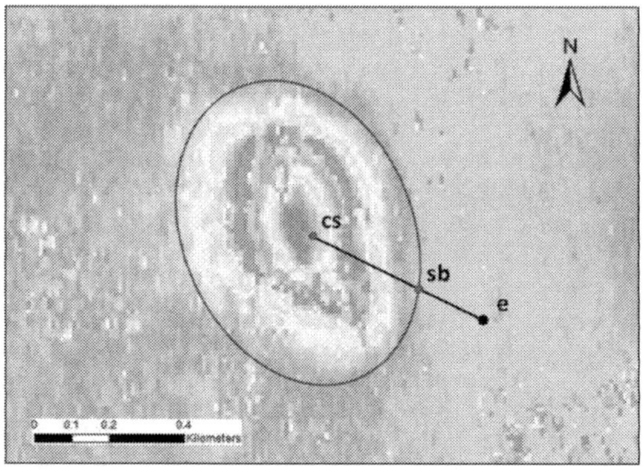

Fig. 3. Proposed geometry method of relation interpretation between centre of subsidence area and epicentre of tremor.

Table 1 presents all the tremors, localised subsidence areas and mines which they refer to. One of the most interesting observations derived from these data is the relationship between interferograms times and the rate of success in finding the subsidence areas. Within 41 interferograms only 6 had temporal baseline (Btemp) of 6 days, the other 25 had Btemp of 12 days or longer. In case of the 6-day repeat cycle, 11 areas were located and also 11 areas where located within the 12-day or longer repeat cycle. This proves 20% better efficiency of the shorter, 6-day repeat cycle. In regard to the correlation between the placement of the epicentre of the tremor and the centre of the subsidence area, only 7 were beyond the subsidence area (within the 22 which were subjected to the analysis). Out of the 13 measured cs-e, 9 were qualified as within the subsidence area and 4 outside.

Table 1. Tremors data collation.

Mining concession area (OG)	[M]	Date of tremor [yy-mm-dd]	Interferometric time base [days]	Image of subsidence	Tremor in subsidence area	Distance cs-e [m]
Libiąż IV	3.8	15-09-30	12	NO		
	3.8	15-11-18	12	NO		
	3.7	15-10-20	12	NO		
	3.1	15-10-10	12	NO		
	2.5	16-11-18	6	YES	YES	220
	2.5	16-11-18	6	YES	YES	230
Bytom III	3.5	16-06-03	12	YES	YES	
Lędziny I	3.4	15-09-15	12	NO		
	3.0	15-09-25	12	YES		
	3.0	15-10-08	12	NO		
	2.9	15-09-08	12	NO		
	2.9	15-11-24	12	NO		
	2.9	17-01-14	6	NO		
	2.5	16-12-22	6	YES	YES	210
Rydułtowy I	3.3	16-06-21	24	YES	YES	300
	3.1	16-12-01	6	NO		
	3.1	16-12-01	6	NO		
	3.1	16-12-01	12	NO		
	3.0	16-08-19	12	YES	NO	
	2.9	15-09-11	12	NO		
	2.8	15-09-08	12	YES	NO	400
	2.5	15-11-18	12	NO		
Radlin I	2.5	16-08-18	12	YES	YES	140
Szczygłowice	3.0	16-04-27	12	NO		
Bieruń II	3.0	16-11-03	6	YES	NO	350
	2.9	16-04-30	12	YES	YES	100
	2.9	16-05-17	12	NO		
	2.9	16-11-16	6	YES	NO	260
	2.9	16-11-16	6	YES	NO	280
	2.9	16-11-21	6	YES	YES	90
	2.9	16-11-21	6	YES	YES	130
	2.7	16-12-13	12	YES	YES	210
	2.7	16-12-13	6	NO		
	2.5	15-09-22	12	YES		
	2.6	15-10-17	12	YES		
	2.5	16-12-01	6	YES		
Bielszowice	3.0	17-01-21	6	NO		
Zabrze I	2.9	16-12-20	6	YES	NO	
	2.5	15-09-17	12	YES	NO	
Niedobczyce	2.8	16-12-09	6	NO		
	2.8	16-12-09	12	YES		

The average distance for these 4 instances was 322.5 m and the average cs-e for the other 9 was at 181 m. For 4 outside events, distances between the edge of the subsidence area and both centres were measured. The edge of the subsidence area identified on interferograms was named sb (for the purpose of this study). This indicates two new definitions: cs-sb – the distance between the centre of the subsidence area and its edge, and sb-e – the distance between the edge of the subsidence area and the epicentre of the tremor. Further analysis showed that only in the area of the OG Rydułtowy I a tremor of 2.80 M magnitude can be qualified as mining-tectonic since cs-e is 400 m, cs-sb is 250 m and sb-e is 150 m. In case of the other three tremors, sb-e distance was only 10 to 30 m and its value limits interpretation – making their origin undefined.

Based on this research, we can divide all tremors into two main categories – tremors inside subsidence area and outside subsidence area.

5 Conclusion

These studies proved that the InSAR technology can be used to divide tremors induced by mining into two categories. The first category tremors are within the border of the six days subsidence trough and the second category is outside the border of the subsidence. However, it is important to mention at this point, that these categories cannot be directly linked to the tremor types classification done by seismologists.

The seismology definition of the first type of mining tremors defines the existence of the direct influence between the underground mining operations and the occurrence of the mining tremor [6]. The underground mining operations are determined by defining the current position of the face of excavation at the moment of the mining tremor occurrence, and the measure of these is the distance between the tremor epicentre and the face of excavation. However, the definition does not indicate how to determine the border distance for which the mining tremor can be considered as directly related to the face of excavation and for which cannot.

Using radar interferometry, it is possible to determine the border of the direct influence of the underground mining operations to the terrain surface, which appeared on this surface during 6 days of the excavation. If the mining tremor occurs within these 6 days, 3 locations of the face of excavation should be considered. The first day is the day of the first SAR image. The next position of the face of excavation shall be marked on the day of the mining tremor and the third one is marked on the day of the second SAR imaging.

The 6-day progress of the face of excavation generates the rock mass deformation whose range is illustrated on the interferogram in the form of the subsidence trough. The external border of the subsidence trough is thus the border of the direct influences of the underground mining operations. This allows us to assume that the epicentres of the mining tremors registered outside this border are not of the first type because they are not directly related to the mining works in face of excavation. The proposed method of classifying categories of the mining tremors, in relation to the definition of tremor types, can significantly simplify the interpretation of the mining tremors.

Of course, during the interpretation of tremors, the distance between the epicentre of tremor (e), the subsidence border (sb) and the centre of the subsidence (cs) should also be taken into account. For example, if a tremor occur with the cs - sb distance being the same or smaller than the sb - e distance, we can be sure that the tremor can be treated as the second type. On the other hand, if the sb - e distance is from 1 m to about 50 m, it does not determine the type of tremor. This is due to the fairly low accuracy of the location of the tremor epicentre. However, it requires further studies on that topic.

The next, more general conclusion based on classical observation on leveling lines [2] and interferometry research, shows that the mining induced tremors (the first type) and the subsidence should be treated as one integrated process of terrain and rock mass deformation.

Another problem, which should be taken into consideration, is an interval in days between the SAR image acquisitions which reached the 6-day repeat cycle with the Sentinel satellite constellation. In the USCB condition, the mining tremors with the magnitude bigger than 2 M occur even every 2 days. Of course, most of them occur in different locations. In the analysed period, there were not any two tremors in the same location within 6 days. However, the emergence of this situation cannot be excluded in the future. Therefore, in the future, a matter of interest should be in obtaining SAR data that have even shorter repeat cycle, for example 2-3 days..

Despite difficulties that occur during acquisitions of SAR data (Btemp 12 days with respect to 6 days) and the coherence of interferograms for the studied subsidence areas, it was proven that this method could be used for monitoring the surface consequences caused by the underground coal mines. The visible correlation between the tremor epicentres and the centre of the subsidence area indicates the need for further scientific exploration of this subject. It should be mentioned that the launching of Sentinel-1B to the orbit significantly improved the data availability for this study.

Based on the described research, the subsidence areas for the further analysis were selected. Best results, in regard to quality of interferograms, were recorded for the OG Bieruń II area. This area is known for its susceptibility to tremors and also the land cover is favourable for the SAR interferometry, which can bring potential economic profits for such areas [16]. The next step of this study will be gathering geological, geotechnical and detailed mining data which allows to analyse the epicentre of the tremor in regard to excavation front movements. This analysis will also include the lithological and tectonic model correlation with the tremors occurrence.

Special thanks to the dr. Z. Perski (PGI-NRI) for all the necessary information's, constructive criticism and suggestions.

References

1. K. Stec, Geophys J Int **168(2)**, 757–768 (2007)
2. E. Popiołek, J. Ostrowski, J. Czaja, J. Mazur, *10th FIG International Symposium on Deformation Measurements,* 77 (Orange, California, USA, 2001)
3. A. Krawczyk Z. Perski, *11th International Congress of the International Society for Mine Surveying* **2**, 209 (Cracow, Poland, 2000)
4. E. Popiołek, H. Marcak, A. Krawczyk, *Warsztaty 2006: Natural hazards in mining,* 339 (2006)
5. K. Mirek,, J. Mirek, Polish Journal of Environmental Studies **20 (4A)**, 253 (2011)
6. J. Gibowicz, A. Kijko, *An Introduction to Mining Seismology* (Academic Press Inc., San Diego, USA 1994)
7. ESA, https://directory.eoportal.org/ (March 2017)
8. USSN, http://www.grss.gig.eu/en/ (March 2017)
9. K. Mirek, Polish Journal of Environmental Studies **18 (3A)**, (2009)
10. X. Li, D. Liu, H. Song, R. Chen, H. Li, Journal of Electronics **31 (5)**, 441 (2014)
11. A.K. Gabriel, R.M. Goldstein, H.A. Zebker, Journal of Geophysical Research: Solid Earth and Planets **94(B7)**, 9183 (1989)
12. Z. Perski., D. Jura, Earth Observation Quarterly **63**, 25 (1999)
13. K. Mirek,, J. Mirek Polish Jour. of Environ. Stud. **25 (5A)**, 57 (2016)
14. M. Przyłucka, G. Herrera, M. Graniczny, D. Colombo, M. Béjar-Pizarro, Remote Sensing **7 (5)**, 5300 (2015)
15. S. Porzycka-Strzelczyk, H. Malik, J. Strzelczyk, *SGEM 2015 Conference Proceedings* **1**,1075 (2015)
16. K. Przeździcki, J. Zawadzki, Systemy wspomagania w inżynierii produkcji **1 (13)**, 11 (2016)

https://doi.org/10.1051/e3sconf/20182600004

Elementary analysis and energetic potential of the municipal sewage sludges from the Gdańsk and Kościerzyna WWTPs

Arkadiusz Ostojski[1,*]

[1] Gdańsk University of Technology, Faculty of Civil & Environmental Engineering, Narutowicza St. 11/12, 80-233 Gdańsk, Poland

Abstract. This paper aims to present municipal sewage sludge (MSS) elementary analysis and energetic potential based on measurement of heat of combustion (higher heating value HHV) and calculation of calorific values (lower heating value LHV). The analysis takes into the consideration water content in sewage sludge, at different utilization stages, in wastewater treatment plants in Gdańsk Wschód and Kościerzyna – Pomeranian Voivodeship. The study yielded the following results (in % dry matter): ash 19÷31 %, C - 31÷36 %, H - 5÷6 %, N - 4÷6 %, O - 28÷32 %, S – 1 %. Calorific value of stabilized sludges in Gdańsk was on average 13.8÷15 MJ/kg. In case of sludges not undergoing digestion from Kościerzyna WWTP, the calorific value was at the level of 17.5 MJ/kg. Thus, sewage sludges are good energy carriers. High water content though is the problem, as it lowers the useful effect of heat. There is no alternative for thermal sewage sludge neutralization, which is in conformity with valid Polish National Waste Management Plan (KPGO 2022).

1 Introduction

Sludge production is directly connected to the sewage treatment. Both, primary and secondary sludges, easily become putrescent. They contain bacteria, pathogenic parasites and exhibit a distinctive feature in the form of repulsive odour. Esthetic, epidemiological, and legal issues make their safe management necessary. Possibilities of sludge processing (management) are the result of:

• physical properties (humidity, dehydration susceptibility, calorific value, heat of combustion), chemical properties (heavy metal contents, dioxins, furans) and sanitary properties;
• economical aspects – accepted cost level of processing and utilization;
• legal aspects.

The basic assumption in waste management, apart from the overriding occurrence prevention or minimization of their quantities, should be resource recirculation. In case of sludges the principle can be fulfilled in two ways: by mass or energy recirculation.

In the first instance it means repeated utilization of organic matter and mineral elements (mostly phosphorus) contained in sludges, which can be achieved by their use in fertilization or land reclamation. Whereas energy recirculation can be carried out by sewage sludge combustion.

According to Polish law and the directive of the European Parliament and Council 2006/12/EEC dated April 5 2006 dealing with waste, waste storage is the last and the least required way of waste processing [1]. The effect of Poland's accession to the EU is the systematic reduction of quantities of sewage sludges utilized in this way (Table 1).

According to regulations of Polish Minister of Economy dated from July 16th 2015 dealing with criteria and procedures permitting wastes to be disposed in given type landfill sites [16] municipal waste exceeding the following values can not be stored from January 1st 2016:

• total organic carbon (TOC) > 5% d.m.,
• ignition losses > 8% d.m.,
• heat of combustion > 6MJ/kg d.m.

Table 1. MSS produced in years 2011-2014 in Poland [5], 2015 [11].

Years	Sludge produced and management within a year in thousands Mg d.m.							
	Total	used				landfilled		sludge accumulated in WWTP
		in land reclamation	in agriculture	in plants cultivation, production of compost	thermally transformed (incineration)	totally	% produced	
2011	519.2	54.4	116.2	31.0	41.6	51.4	9.89	212.4
2012	533.3	50.3	115.0	33.3	56.6	46.8	8.77	208.1
2013	540.3	29.4	105.4	32.6	72.9	31.4	5.81	219.8
2014	556.0	22.0	107.2	46.3	84.2	31.5	5.67	226.0
2015	568.0	19.2	107.5	47.1	79.3	40.5	7.13	246.9

* Corresponding author: arkusost@pg.gda.pl

To improve waste management operation in Poland the National Waste Management Plan is prepared every four years (KPGO). The latest National Waste Management Plan 2022 was prepared in 2016 in accordance with the new framework directive 2008/98/EEC dealing with waste [2], which replaced the directive 2006/12/EEC. The plan KPGO 2022 in context of sewage sludges underlines *'limited possibilities of prevention in occurrence of municipal sewage sludges'* at the same time anticipating the rise of produced sludges. Among the basic assumptions of sludge management for the coming years there are: restrictions in their storage, increased quantities of municipal sewage sludges, which will be thermally processed, maximal utilization of biogenic substances contained in sludges with fulfillment of all requirements referring to sanitary, chemical and environmental safety conditions.

Quality requirements, in particular related with low concentration of heavy metals in non-industrious sludge use in treatment plants for agricultural use, are very strict and difficult to achieve in municipal and industrial catchment areas. Yet, the basic obstacle in agricultural use results from quantity restrictions of applied annual sludge doses. The order of Minister of Environment dated from February 6th 2015 dealing with municipal sewage sludge [15] limits the annual dose of municipal sewage sludge in agriculture to 3 tons of dry matter (d.m.) per hectare, and in case of single use within two or three years up to 6 Mg/ha/2 years and 9 Mg/ha/3 years. As a result big municipal treatment plants with population over 100 thousand, with the equivalent number of inhabitants (ENI) for complete agricultural sludge management per year, would need over a few hundred hectares of arable land for complete agricultural sewage sludge management to be used once within 3 years. It is clear that providing such enormous acreage is impossible.

Nevertheless, it is absolutely necessary to remove harmful organic matter containing heavy metals. In case of exceeding concentration of heavy metals or sanitary indicators, instead of mass recirculation it is an obligation to recover energy by sludge combustion or co-combustion. Basing on the official programmes accepted by Poland's government (KPGO 2022) thermal methods of sludge processing are to be recognized as the most promising and developmental. The purpose of the carried out research of heat combustion and elementary analyses of sewage sludges in the two largest treatment plants in the Pomeranian region was to examine their energy potential. The composition of sludges before and after fermentation allows to assess the correctness of the conducted stabilization process. HHV and LHV of sludges may be decisive in the choice of thermal treatment method, parameters of the process, but also in earlier management technology, for instance the intensity of biogas production, dehydrating methods. They may also be decisive in profitability of the whole undertaking of thermal utilization.

2 Experimental - investigation sites

2.1 The WWTP in Gdańsk Wschód

Gdańsk Wschód is the largest wastewater treatment plant in the Pomeranian Province, it collects municipal sewage from Gdańsk, Sopot, Straszyn and the surrounding districts: Pruszcz Gdański, Kolbudy, Żukowo. The person equivalents (p.e.) in 2009 was 650 000, the daily average flow was almost 94 thousand m^3/day.

The treatment technology includes mechanical, biological treatment and sludge processing. Sewage after treatment is directed to the waters of Gdansk Bay by pipeline of 2.5 km length. The mechanical treatment system consists of 4 mechanical screens, 3 grit chambers and 4 primary clarifiers. Mechanically treated sewage is then pumped to the separating chamber and later to one of the six parallel biologically treated process lines. Biological treatment, where there are integrated processes of removal of organic carbon, nitrogen and phosphorus compounds carried out, is based on MUCT – Modified University of Cape Town System. Each of the six bioreactors consists of five chambers (dephosphatation, denitrification of 1st and 2nd degree, nitrification and deoxidation), which assure keeping differentiated oxygen and hydraulic conditions.

The treated sewage from biological reactors is carried to two radial secondary clarifiers (there are 12 of them), which allow separation of activated sludge from sewage. Then sewage is carried by the open channel to the measuring flume, and later by closed channel to the lift station to the Gdańsk Bay by pipeline. Gravitationally thickened sludge from the lower part of secondary clarifiers is partly recirculated to the first hypoxemic chamber of the biological reactor and partly to mechanical sludge thickening (filter presses). Then it goes to the mixing chamber and together with preliminary sludge it goes to closed fermentation chambers - the place of sludge collection before fermentation. After anaerobic methane fermentation sludge is mechanically dewatered in sedimentation centrifugal separators – samples are collected from the pipeline carrying fermented sludge to separators and from the screw conveyor behind the separator. The average sludge production per day is over 31 Mg d.m.

2.2 The WWTP in Kościerzyna

The municipal wastewater treatment plant is located in the southern part of the town. Its present condition is the effect of the modernization and expansion terminated in September 2007. The new mechanical and biological treatment plant consists of two completely independent technological treatment lines using activated sludge method in A2/O technology with increased removal of nutrients. Raw sewage is delivered to the wastewater treatment plant from the storage tanks, which makes their parameters average. In the course of the extension trietary treatment was applied, ie. elementary filtration of secondary effluent WWTP, which makes full retention and averaged sewage possible , as well as using them as technological waters in the treatment process. The modernized wastewater treatment plant collects sewage from 33.8 thousand person equivalents (p.e.), the average flow capacity of the modernized waster water treatment plant was 3,379.4 m^3/d in 2009 and the maximal flow for rainy weather was 12,000 m^3/d. The treatment plant delivers treated sewage to the Bibrowa River and then to Wierzysko Lake and the Wierzyca River. The treatment plant has its own waste management, which consists of the mixing facility where the mechanically dewatered sludge in presses is mixed with straw and then directed to the closed chambers of the compost site. The daily average volume of produced sludge in 2008 was about 2 Mg d.m. (742 Mg s.m./year).

3 Material and methods

The samples of volume from 1 to 3 liters, after collection were placed in cooler boxes in 4 °C. The first introductory determinations were started on the day of collection or they were stored in the refrigerator till the next day in the laboratory of Faculty of Civil and Environmental Engineering, Gdańsk University of Technology. After drying and determining humidity the samples were transferred to the laboratory of Faculty of Chemistry to indicate heat of combustion, calorific value and sulphur content. Then small quantities of a few grams of dried sludge were delivered for elementary analysis to the laboratory in Molecular and Macromolecular Research Centre of Polish Academy of Science in Łódź.

From the municipal waste treatment plant in Gdańsk 123 averaged samples were collected (mixed primary sludge and secondary sludge for fermentation – 41 samples, fermented sludge – 42 samples, dewatered sludge in the centrifuge – 40 samples), WWTP in Kościerzyna – 68 averaged samples (dewatered sludge in presses – 38 samples and composted sludge – 30 samples). Totally, in the period of over 2 years 191 samples were collected from both WWTPs.

3.1 Analytical procedures

The analyses were conducted according to PN-G-04571 "Solid fuels. Determination of C, H and N by automatic analyzers. The Macro method". Although the standard concerns only the determination of carbon, hydrogen and nitrogen, the determination of sulphur was carried out in a similar way.

The determination of carbon, hydrogen, nitrogen and sulphur was carried out by the Vario MICRO cube automatic analyser produced by Elementar Analysensysteme GmbH. As a result of an exothermic reaction it is burnt in oxygen in 1150 °C and the burning products N_2, CO_2, H_2O, SO_2 are adsorbed in a special column. Next, according to the program, they are desorbed at the right temperature and are directed to a thermal conductivity detector (N_2, CO_2, H_2O). In this case the SO_2 to an infrared detector. The analyser is connected to the computer, which both controls the combustion process and calculates the content of the elements in the analysed sample. The determination was conducted in two test portions of each substance.

The determination of oxygen was carried out using the automatic analyser EA1108 produced by Carlo Erba. The test portion is weighed in a silver foil container and placed inside the incinerator heated up to 1070 °C, where it undergoes pyrolysis. The quantitative conversion of oxygen to carbon oxide takes place in nickel-coated carbon. Gases are directed to a chromatographic column, and then to a thermal conductivity detector. By means of the Eager 200 program, both the calibration curve and the results of the analysis are obtained. As in the case of the analyses mentioned earlier, the determination was carried out in two test portions of each substance.

The analysis of the energetic potential carried in the sludge from the wastewater treatment plants in the Pomeranian Voivodeship based on measurement of heat of combustion (higher heating value HHV) and calculation of calorific values (lower heating value LHV - including moisture content) was conducted using the automatic calorimeter KL–10. The apparatus measures the heat of combustion of solid fuels, such as: peat, hard coal, brown coal, hard coal briquettes, brown coal briquettes, coke, semi-coke as well as inexplosive flammable organic matter. A sample of fuel is completely burnt in oxygen under pressure in a bomb calorimeter immersed in water, where increase in water temperature is measured. The heat of combustion of the fuel is automatically calculated and shown on a digital display. The calorimeter measures heat exchange, and characteristic temperatures converted to digital form are stored by the system. Heat gain is measured by a thermistor sensor.

Heat of combustion is the quantity of heat released during complete combustion of solid fuel in the bomb calorimeter with oxygen under the temperature of 25 °C (PN-81 G-04513, 1981). Calorific value is the heat of combustion reduced by heat of water evaporation emitted during combustion of fuel and created from hydrogen contained in the fuel in MJ/kg (PN-81 G-04513, 1981). Calorific value – it is the useful heat effect of combustion [3].

4 Results and conclusions

The fuel is composed of the combustible substance consisting of chemical compounds of elementary carbon (C), hydrogen (H) and sulphur (S), as well as mineral substances (ash) and moisture. Fuel can be used for external and internal combustion . Depending on the state of aggregation one can distinguish solid, liquid and gaseous fuel, and depending on the origin there is natural, artificial and waste fuel [17, 18].

Solid fuels are characterized by a series of indicators, such as heat of combustion HHV (MJ/kg), calorific value LHV (MJ/kg), ash content (%), sulphur content (%). Fuel's calorific value and heat of combustion depends on its chemical composition disclosed in the result of technical analysis (moisture content, ash content or volatile substances) and elementary analysis (C, H, O, S). The average combustible material of mixed sludge

(60 % primary sludge, 40 % secondary sludge) from German urban agglomerations consists of [13]: C – 50 %, O – 38 %, H – 7 %, N – 4 %, S – 1 %, Cl ≤ 0.2 %, F ≤ 0.01 %.

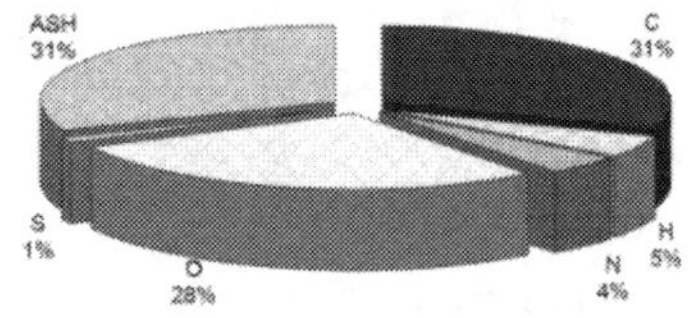

a) in the sludge dry mass (dm*)

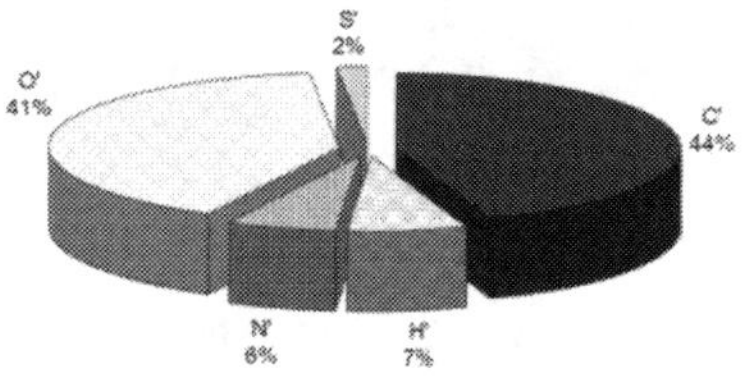

b) in dry organic mass (daf**)

Fig. 1. The average content of C, H, N, O, S in the sludge from the centrifuge in Gdańsk WWTP.

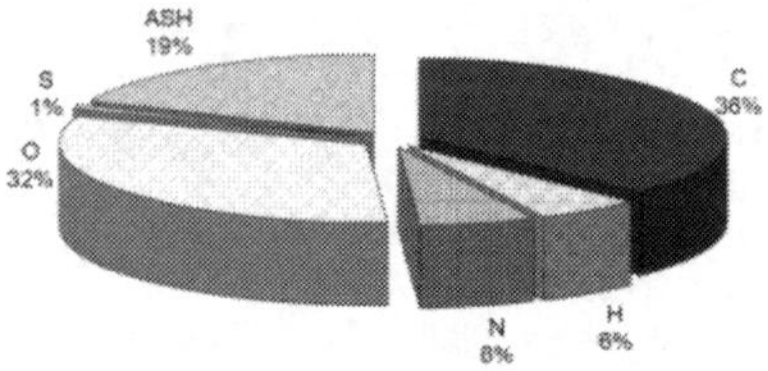

a) in the sludge dry mass (dm*)

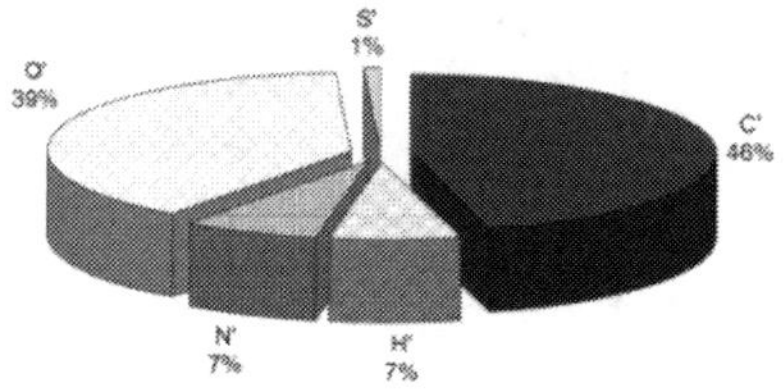

b) in dry organic mass (daf**)

Fig. 2. The average content of C, H, N, O, S in the sludge dewatered in presses in Kościerzyna WWTP.
* dm – dry matter, **daf – dry and ash free matter

Table 2. Elementary analysis and calorific values (LHV, HHV) for the sludges and the coal.

Sample material	Raw sample								
	Water	Ash	LHV	C	H	N	S	O	HHV
	%	%	MJ/kg						MJ/kg
Kościerzyna dewatered sludge in presses	81.3	18.8	-0.048	37.3	5.6	5.9	0.8	31.7	17.44
Kościerzyna composted sludge	65.4	18.8	3.47	37.8	5.1	2.9	0.6	34.8	17.90
Gdańsk before digestion	94.9	23.2	-2.53	36.8	5.3	4.9	0.8	29.0	17.69
Gdańsk digested sludge	95.1	34.9	-2.59	28.8	4.4	4.0	0.9	27.0	13.83
Gdańsk from the centrifuge	79.4	31.3	0.12	30.6	4.6	4.3	1.1	28.1	14.92
sludge from very low industrialized town (Spain) [6]	4.3	31.2		38.2	4.3	4.5	0.9	20.9	17.61
sludge from higher degree of industrialization town (Spain) [6]	3.9	53.8		22.7	3.3	3.1	1.6	15.5	9.48
coal (Spain) [6]	1.2	17.7		73.5	3.2	1.6	2.2	1.8	28.87
coal (PL) [12]	9.8	17.9	2.30	67.1	4.7	1.5	0.5	8.3	
dry sewage sludge (PL) [12]	7.4	36.8	1,19	33.6	4.8	4.8	1.3	18.5	
Soft brown coal: [9]	Raw sample				Dry ash free matter				
Rhineland (D)	50–62	5–20	6.3–9.6	68.3	5	0.5	0.5	27.5	26.38
Helmstedt (D)	42–46	12–22	9.2–10.5	72.6	5.8	0.4	4.4	16.7	29.75
Schwandorf (D)	50–58	6–20	6.3–7.5	63.6	5	1.3	4	26.1	25.33
Lausitz (D)	55–60	2–5	8.2–8.5	67.5	5.2	1	0.8	25.5	25.37
Leipzig (D)	50–55	5–7	9.0–11.0	71.6	6.1	0.7	2.1	19.5	28.35
Halle-Bitterfeld (D)	52–56	5–7	9.6–10.0	72	5.5	0.8	3.4	18.3	29.81
Ptolemais (USA)	52–60	6–22	3.6–6.7	65.3	5.3	1.6	0.5	26.5	25.25
Megalopolis (USA)	60–64	13–17	2.8–4.0	60.5	6.2	1.3	1.4	30.6	24.45
Yallourn (AUS)	63–72	1–2	5.0–7.5	67.5	4.8	0.7	0.3	26.7	25.54
Patnow (PL), Lusatia (D)	52–58	6–15	8.0–8.8	73.6	5.1	0.5	1.1	19.7	28.56
Gyöngyös/Visonta (H)	46–54	15–30	5.0–6.7	63.8	4.8	1.1	3.5	26.8	24.83
Elbistan (TR)	48–62	8–24	3.3–6.2	61.4	5.1	0.8	5.1	29.6	23.69

Despite different sewage flows, various technologies of waste treatment, sludge treatment, and processing the composition of dry organic mass of dewatered sludges (daf) from both municipal wastewater treatment plants did not show any significant differences C – 44÷46 %, H – 7 %, N – 6÷7 %, O – 39÷41 %, S – 1÷2 % (Fig. 1b and 2b). Comparing the achieved results from Gdańsk and Kościerzyna with German literature data [13] lower carbon content (6% on average) and higher oxygen content (3% on average) should be underlined, however the differences are small. Greater differences in carbon content are to be noticed when comparing the obtained results with sludge composition coming from Danish WWTPs: C – 52÷58 %, H – 7÷8 %, O – 29÷31 %, N – 4÷9 %, S – 1÷1.5 % [8], Czech WWTPs: C – 67 %, H – 5%, O – 25 %, N – 2.2 %, S – 0.8 % [10] and Chinese WWTPs: C – 52.4 %, H – 9.3 %, O – 27.4 %, N – 7.9 %, S – 3.0 % [19].

The difference in the applied wastewater treatment technologies and sludge processing is noticed when dry sludge is compared (dm – without omitting ash content) (Fig. 1a and 2a). Sludges from Gdańsk are considerably highly mineralized than sludges from Kościerzyna, which results from methane fermentation before dehydration. Sludges from Kościerzyna are excessively unstabilized. The comparison of percentage composition of sludges from both treatment plants with literature data is presented in Table 2.

The heating value of raw sludge is about 17 MJ/kg; for activated sludge, about 15 MJ/kg, and for stabilized sludge (digested: anaerobic, aerobic, or lime stabilization) around 11 MJ/kg [14]. The value of HHV was considerably higher in excess sludge from Kościerzyna, it was on the level of 17.5 MJ/kg. The fermented sludge from Gdańsk was also characterized by higher value of HHV, which was almost 14 MJ/kg.

Heat of combustion of dry sewage sludges is about 15÷17 MJ/kg and is almost twice lower than heat of combustion of good quality coal.

Generally, it is assumed that one ton of hard bituminous coal is energy-efficient to 2 tons of dry sludges.

The combustion process can be autothermic or it can require heat supply. When one third of the sludges is inflammable, the heat from burning dry sludge does not exceed 14 MJ/kg, and after subtracting the heat necessary to evaporate water, the calorific value of the sludges decreases to about 9 MJ/kg. The calorific value of the digested sludge is lower than that of the raw non-digested sludge by about 2 MJ/kg [7]. If the humidity of sludges is high, the heat of combustion is equal to 0. If the heat from burning sludge is insufficient to evaporate the water present in it, it is necessary to supply an additional heat source e.g. coal, heating oil, straw, biogas or woodchips.

According to Austrian researchers, the calorific value of the wastewater sludge which contains 50 % of organic substance and dewatered till dry matter content 50 % is equal to 4 MJ/kg. The calorific value of the one that contains 75 % of organic substance and dewatered – having a moisture content of 50 % – is equal to 6.5 MJ/kg. When the moisture content is equal to 85 % in the hydrated sludge, which contains 50 % of organic substance, the calorific value is equal to 0 [7].

The values of heat of combustion obtained in the course of laboratory tests (Fig. 3) are at the level 13.8÷17.7 MJ/kg and the variation of this parameter referred to the applied treatment processes. Methane fermentation in Gdańsk caused reduction of heat of combustion by about 3÷4 MJ/kg in comparison to not fermented sludges. In comparison to the literature data fermented sludges from Gdansk waste treatment plants were characterized by high value of HHV and low content of ash – just 34÷35 % at this stage of treatment. Czech sources say that digested sludge contented, on average, 49.2 % of organics and 50.8 % of ash in dry matter. The higher heating value was 12.18 MJ/kg of dry matter. The lower heating value was 11.23 MJ/kg of dry matter [4].

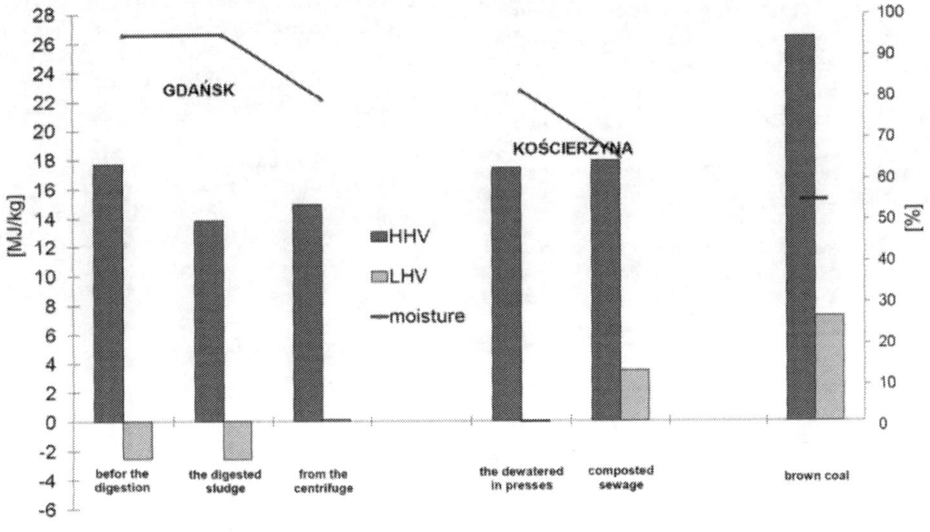

Fig. 3. The average HHV, LHV and moisture of the sludges from WWTPs Gdańsk, Kościerzyna and brown coal.

The achieved values of heat of combustion are consistent with the presented literature data, although fermentation and stabilization processes could be more efficient and reduce the value of heat of combustion even more. In case of Gdańsk it is possible to obtain bigger quantities of biogas. Nevertheless, it should be connected with proper utilization of fuel to make it sensible.

It must be noted that composting, which is the final stage of sludge processing in Kościerzyna WWTP, improves energy parameters of the sludge HHV – 17.9 MJ/kg, LHV – 3.5 MJ/kg with humidity > 65 %. Composting prepares sludge for natural utilization and does not exclude thermal methods of utilization.

5 Conclusions

1. Although it is consistent with the matter cycle in nature, utilization of sludges for natural purposes would be more and more difficult. Requirements for treated sludges are higher and higher, there is no sufficient acreage for fertilization or reclamation – particularly in large WWTPs. However, natural methods should be promoted for small treatment plants collecting mainly municipal sewage. Energy utilization of sludges seems to be attractive provided combustion gas is properly treated.
2. Comparing the results of the elementary components of the sludges with the literature, a lower content of carbon and higher content of oxygen can be observed.
3. Sludges are a good source of energy. The average heat of combustion (HHV) of wastewater sludge was 13.8÷15 MJ/kg for the digested sludge and 17.5÷17.7 MJ/kg for the non-digested sludge from the both NBR (nutrients biological removal) WWTP. Two - three tons of sludges, after drying of about 40% humidity, are the equivalent of one ton of hard bituminous coal. Thus, sewage sludges are a good energy carrier. High water content is the problem, as it lowers the useful effect of heat.
4. Composting prepares sludge for natural utilization but also improves energy parameters of the sludge (HHV – 17.9 MJ/kg, LHV – 3.5 MJ/kg with humidity >65 %) and does not exclude thermal methods of utilization.

Financial support from Island, Liechtenstein and Norway through the EEA Financial Mechanism and Polish Ministry of Science and Higher Education (project no PL0085) is gratefully acknowledged.

References

1. *Directive 2006/12/EC of the European Parliament and of the Council of 5 April 2006 on waste* (OJ EU L 114 (49), 2006)
2. *Directive 2008/98/EC of the European Parliament and of the Council of 19 November 2008 on waste and repealing certain Directives*, (OJ EU L 312(51), 2008)
3. Z. Grabowski, J. Oleszkiewicz, *Sludge combustion. Materials from International Training Seminar 'Theoretical basic and practice of sludge processing and management.* (Ed. LEM, Cracow, Poland, 1998)
4. L. Houdkova, J. Boran, V. Ucekaj, T. Elsaser, P. Stehlik, Thermal Engineering **28 (16)**, 2083 (2008)
5. *Polish National Waste Management Plan 2022* (Official Journal of the Republic of Poland - Monitor Polski 2016/784, 2016)
6. M. Otero, z M. E. Sánche, A. I. García, A. Morán, Journal of Thermal Analysis and Calorimetry **86 (2)**, 489 (2006)
7. A. Schubering, Technical-Research Symposium *The newest technology for Environment protection in Austria*, copied typescript (Warsaw, 1992)
8. N. Simonsen, C. Bisgaard, B. Nielsen, 10th EWPCA Symposium *Sludge Treatment and Reuse,* manuscript (IFAT '96, Krueger A/S Technical Division, Denmark, Munich, 1996)
9. H. Spliethoff, *Power Generation from Solid Fuels*. Springer Heidelberg Dordrecht London New York, ISBN 978-3-642-02855-7 (2010)
10. P. Stasta, J. Boran, L. Bemar, P. Stehlik, J. Oral, Thermal Engineering **26 (13)**, 1420 (2006)
11. *Statistical Yearbook of the Republic of Poland* (Central Statistical Office, 2016)
12. S. Stelmach, R. Wasilewski, J Mater Cycles Waste Manag **10 (2)**, 110 (2008)
13. S. Solmaz, Korrespondenz Abwasser **45(10)**, (1998); Gaz, Woda i Technika Sanitarna **III/99**, Annex 1-8 (1999)
14. *The National Municipal Wastewater Treatment Program (KPOŚK)* (Regulation of the Ministry of Environment, Republic of Poland, Warsaw, 2003)
15. *The order of Minister of Environment dated from February 6th 2015 dealing with municipal sewage sludge* (Journal of Laws 2015/257, 2015)
16. *The order of Minister of Economy and Labour dated from July 16th 2015 dealing with criteria and procedures permitting wastes to be disposed in given type landfill sites* (Journal of Laws 2015/1277, 2015)
17. J.W. Wandrasz, 4th Conference *Improvement of energy and environment utilization* **I**, 97 (ODKT RS NOT, Warsaw, 1996) (in Polish)
18. J.W. Wandrasz, A.J. Wandrasz, *Formed fuel. Bio-fuels and refused derived fuels* (RDF) (Publishing House 'Seidel-Przywecki' Sp. z o.o., Warsaw, 2006) (in Polish)
19. H. Xiao, X. Ma, Z. Lai, Applied Energy **86 (9)**, 1741 (2009)

The Optimum Dataset method – examples of the application

Wioleta Błaszczak-Bąk[1], *Anna* Sobieraj-Żłobińska[2,*], and *Beata* Wieczorek[3]

[1]University of Warmia and Mazury in Olsztyn, Faculty of Geodesy, Geospatial and Civil Engineering, Institute of Geodesy, Oczapowski St. 1, Olsztyn, Poland
[2]Gdańsk University of Technology, Faculty of Civil and Environmental Engineering, Department of Geodesy, Narutowicza St. 11/12, Gdańsk, Poland
[3]University of Warmia and Mazury in Olsztyn, Faculty of Geodesy, Geospatial and Civil Engineering, Department of Land Surveying and Geomatics, Heweliusza St. 12, Olsztyn, Poland

Abstract. Data reduction is a procedure to decrease the dataset in order to make their analysis more effective and easier. Reduction of the dataset is an issue that requires proper planning, so after reduction it meets all the user's expectations. Evidently, it is better if the result is an optimal solution in terms of adopted criteria. Within reduction methods, which provide the optimal solution there is the Optimum Dataset method (OptD) proposed by Błaszczak-Bąk (2016). The paper presents the application of this method for different datasets from LiDAR and the possibility of using the method for various purposes of the study. The following reduced datasets were presented: (a) measurement of Sielska street in Olsztyn (Airbrone Laser Scanning data – ALS data), (b) measurement of the bas-relief that is on the building in Gdańsk (Terrestrial Laser Scanning data – TLS data), (c) dataset from Biebrza river measurment (TLS data).

1 Introduction

Rapidly developing measurement technologies such as LiDAR (Light Detection And Ranging), MBES (Multi Beam Echo Sounder), bring large amounts of data. With such datasets, digital models of measured objects can be generated, for example: Digital Terrain Model or the Digital Elevation Model. However processing of such amount of data, especially if there are real-time changes occurring, is virtually impossible or very difficult [14]. Constantly, there are ongoing works on improving the processing of large datasets. Such datasets can be used to build the Spatial Information System (SIS), as well as being the source of many studies. One of the projects that aims to provide an advanced use of spatial data is the Center for Spatial Analysis of Public Administration (Centrum Analiz Przestrzennych Administracji Publicznej - CAPAP). Within the project tasks and tools related to spatial data processing are planned. The project is planned to include execution of tasks related to tools enabling the use of spatial data. One of such tool is an analytical platform that enables advanced spatial analysis, including 3D data analysis, as well as interpretation and visualization of the results in textual and graphical form.

In this and many other cases [e.g.: 1, 16] it is necessary to reduce the the measurement set. Decrease of the data can be conducted by means of reduction or generation. Generation is decreasing the dataset by creating a grid. In this method we have new points instead of points with the original coordinates [2, 11]. While reduction decreases the dataset by removing some points according to the given algorithm, the remaining points are original points from the measurement [4, 5, 6, 9]. For people using data in the form of point clouds, it is better and easier to use real data. Therefore, reduction is a better option.

Reduction of the dataset is a problem that needs to be properly planned so that the dataset after the reduction meets all the user's expectations. Evidently, it is best if the result is the optimal solution for the adopted criteria. It can be achieved by using the Optimum Dataset (OptD) method [7, 8] - the optimum method of reduction. This paper presents various variants of this method's applications. The obtained reduced datasets were generated on the basis of data from: (a) measurement of the Sielska street in Olsztyn (Airborne Laser Scanning data – ALS data), (b) measurement of the bas-relief that is on the building in Gdańsk (Terrestrial Laser Scanning data – TLS data), (c) dataset from Biebrza river measurement (TLS data).

2 The Optimum Dataset method

The OptD method is designed to reduce large sets of data such as ALS data and TLS data. The method takes into account the different levels of reduction in the individual parts of the processing area. And as a result, there are more points in detailed part of scanned object. In the case of uncomplicated structures or areas, the number of points is much less. Only those points that are significant will remain, and the generated model will meet the predetermined parameters, for example, the accuracy of the obtained model. The method has been described in details and presented in [7, 8].

* Corresponding author: annsobie@pg.edu.pl

https://doi.org/10.1051/e3sconf/20182600005

The most important step in the OptD method is the selection of a line generalization algorithm that creates points in the measurement strips with the coordinates (x_i, z_i) for vertical strips or with the coordinates (y_i, z_i) for horizontal strips. Research shows that it is enough to use the generalization algorithm in the chosen system. There is no need to create vertical and horizontal strips at the same time.

During the reduction, the position of the points relative to each other is considered. The selected algorithms presented for example in [10, 13, 18] determines the degree of reduction by applying the appropriate tolerance range.

3 Examples of applications

3.1. DTM generation

Digital Terrain Model DTM is one of the most popular product which can be used in many fields of economy and science. In order to created it a reliable data is necessary. One of the source for DTM generation is point clouds obtained during laser scanning measurements.

The study area for which airborne laser scanning measurement was conducted is a fragment of the national road No. 16, a Sielska street in Olsztyn, located in the Warmia-Mazury. Measurement made by Visimind Ltd enabled the acquisition of point clouds. A fragment of the original ALS dataset (144500 points), which was used as a study area of this research, is presented in Fig. 1a.

The selected fragment was filtered by using adaptive TIN model method implemented in own software. As a result of the filtration, there are two sets of data: a) the set of points showing the topography (108313 points), b) a set of points showing the details points (36187 points). Point cloud after filtration, which was used to generate the DTM is shown in Fig. 1b.

Point cloud after filtration comprising only ground points was optimized by OptD-single method. Application of the OptD-single method selected the optimum solution, which is presented in Fig. 1c.

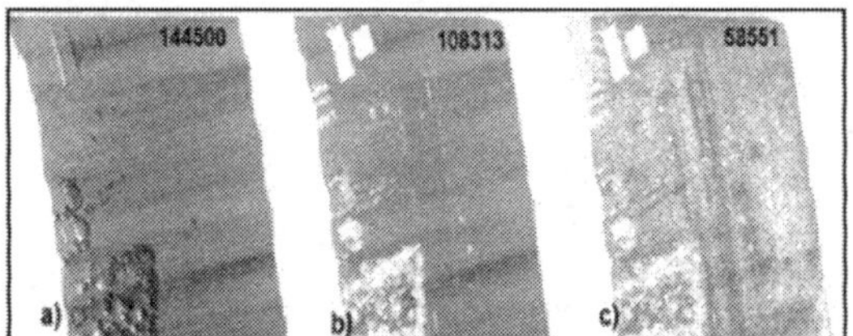

Fig. 1. Fragment of ALS point cloud original, b) after filtration, c) after the OptD-single method was applied (source: Błaszczak-Bąk 2016).

Fig. 2. DTMs generated on the basis of: the original set consisting of 108313 points (left), the set decreased by means of the OptD-single method consisting of 58551 points (right) (source: Błaszczak-Bąk 2016).

Standard deviation SD and coefficient of determination D (denoted also as R^2) were calculated for the generated DTMs. Coefficient of determination D is the measure of model adjustment (the closer to 1, the better the match of the model to another). The results are presented in Table 1.

Table 1. Parameters of generated DTMs.

Parameters	DTM based on original data	DTM based on data decreased by OptD-single method
SD [m]	0.071	0.083
D	-	0.980

The coefficient of determination indicated good adjustment of DTM generated on the basis of reduced dataset to DTM based on original dataset. The SD is higher for DTM for OptD-single method only about 0.01m.

3.2 3D modelling

LiDAR technology makes it possible to obtain point clouds with a high density, resulted from the scanning resolution set by the operator. Issues related to the processing of LiDAR data can be found, among others, in [12], where automatic estimation of agricultural tree geometric parameters and its accuracy are presented, in [14] who described the processing of 3D modelling triangle meshing. In paper [3] the authors discussed the problem of 3D modelling of terrestrial objects. In all of these papers the whole point cloud was used and the density of points was relatively the same throughout the scanned object. Very often such amount of data and uniform density is not needed, especially when the main focus is on the features of relatively flat object (e.g.: bas-relief) [2]. For this purpose, it is worthwhile to use the OptD method [7]. It allows to reduce the obtained dataset without losing the characteristic points and in result the density of points is various throughout the scanned area. Application of the OptD method, especially for objects with complex structure, can help to reduce the number of points without losing the information necessary for proper modelling.

As a research facility, the building located within the University of Gdansk was used. On the front elevation (north wall) there is a bas-relief. It shows the image of a

https://doi.org/10.1051/e3sconf/20182600005

dragon and a man coal-stoker with a shovel. The fragment of obtained point cloud including the bas-relief consists of 753583 points. The whole bas-relief is shown in Fig. 3.

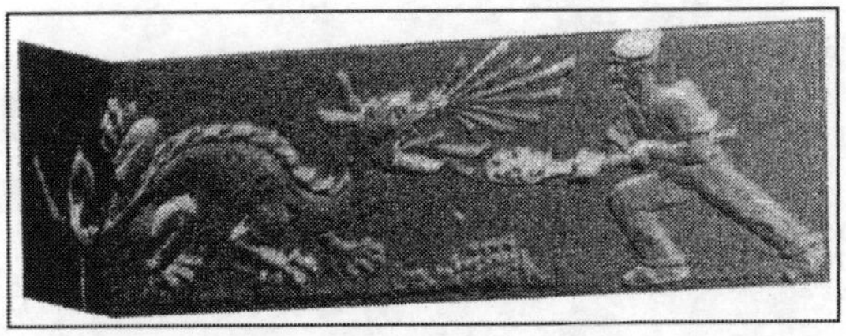

Fig. 3. Point cloud with bas-relief (source: own study in the CloudCompare v.2.7.0).

TLS point cloud with bas-relief was processed by means of the OptD-single method. After optimization 488715 points remained. It is about 64 % of the points of the original point cloud.

Fig. 4 presents a bas-relief with lower density of points, however a dragon and a coal-stoker with a shovel are still very well visible. The characteristics details are still preserved.

Fig. 4. Point cloud after reduction (source: own study in the CloudCompare v.2.7.0).

Decreased dataset was used to build 3D model mesh. A mesh is a set of vertices, edges and faces that defines the shape of a polyhedral object in 3D computer graphics. The faces usually consist of triangles (triangle mesh), quadrilaterals, or other simple convex polygons, since this simplifies rendering, but may also be composed of more general concave polygons, or polygons with holes [17].

On the basis of original TLS dataset and dataset after application of the OptD method the 3D model were generated. In Fig. 5a and 5b original and reverse views of the bas-relief are presented. The reverse view was chosen to use because the details of bas-relief are much better visible. Models were generated in CloudCompare v.2.7.0 as mesh.

Generated 3D models are almost identical. There are no significant differences. The advantage of 3D model generated on the basis of optimized dataset is the decreased number of points in input set, what in result accelerated the modeling. In the case of decreasing the set about 35.1 %, the time needed for modeling decreased about 33.3 % (360 sec for whole set, 240 sec for optimized set).

Fig. 5. Bas-relief 3D model generated on the basis of optimized dataset a) original view, b) reverse view (source: own study in the CloudCompare v.2.7.0).

In order to compare the obtained results and indicated the differences, the contour models based on GRID were also generated for fragment of the bas-relief (coal-stoker figure). They are presented in Fig. 6a and Fig. 6b.

For generated models OriDset (original dataset) and OptDset (optimized dataset) following assessment parameters were calculated:

- mean value of height difference (Δh_{mean}),
- coefficient of determination (D),
- standard deviation (SD).

Results are presented in Table 2.

Table 2. Comparison of generated GRIDs.

Parameters:	Bas-relief GRID	
	OriDset	OptDset
Δh_{mean} [m]	-0.001	
SD [m]	0.234	0.245
D	-	0.99

Comparison of mentioned parameters shows that the coefficient of determination is very close to the value 1 for the OptDset. Thus, GRID generated from the reduced dataset is identical with GRID generated on the basis of the whole dataset representing the bas-relief, what means that interpolated heights in corresponding vertices are almost the same. Standard deviation SD calculated for OriDset and OptDset differs only about 0.001m, so it is almost negligible difference.

 https://doi.org/10.1051/e3sconf/20182600005

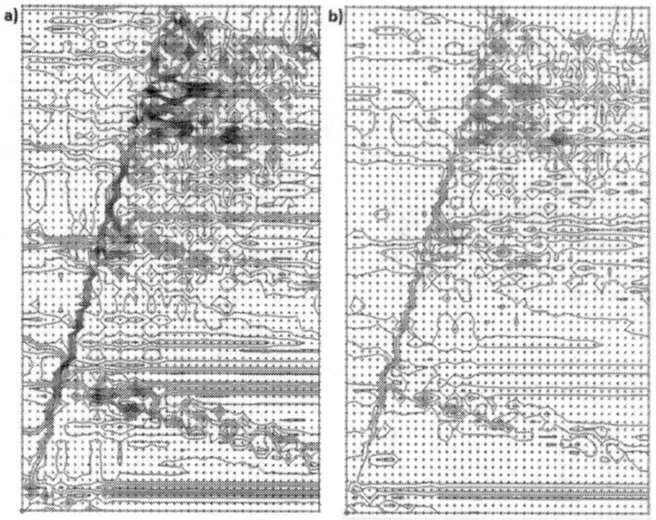

Fig. 6. The contour models for: a) original set, b) optimized set (source: own in the Surfer v.8).

3.3 Point cloud visualization

Point clouds and its visualization are good sources of information in case if there are changes within measured objects/surfaces. One of the example is meandering river area, where flow of inland waters, soil erosion and direct or indirect human activities are causing deformation of the land. Usefulness of spatial visualization of terrain in order to analyze such changes is determined by spatial data and methods of processing and modeling. Both, data and these methods, should allow to obtain a model with fixed or the appropriate geometric accuracy, in particular the altitude. The data and methods also should enables the appropriate mapping surface detail of the terrain model, wherein the amount of data need to be optimal. The large amount of data does not mean that visualization will be better or more accurate. Therefore, it was decided to apply the OptD method on the data acquired during measurement of one of meandering section of Biebrza river near village Goniądz conducted in May 2012. The C10 Leica Geosystem scanner was used.

In Fig. 7 there is a fragment of original TLS point cloud (from one station).

Characteristics of original and decreased sets are presented in Table 3.

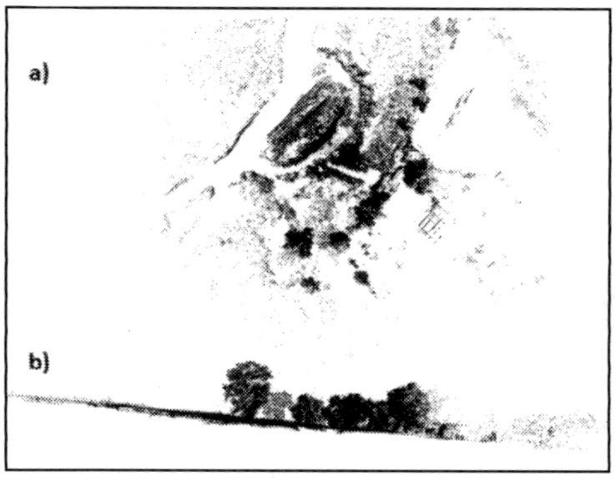

Fig. 7. The fragment of original TLS point cloud from one station for Biebrza river: a) top view, b) side view (source: own study in CloudCompare v.2.7.0).

Table 3. Characteristics of original and decreased sets.

Characteristics	Original set	Decreased set
Number of points	1022857	451915
Length [m]	396.531	396.531
Width [m]	399.279	399.279
Max height [m]	76.714	76.714
Min height [m]	-6.171	-6.171
H standard dev [m]	4.036	5.521
Mean height [m]	-0.826	0.550
SD [m]	4.036	5.521

* Corresponding author: annsobie@pg.edu.pl

The set was processed by means of the OptD-single method. There were 570942 points removed, what is 56 % of the original set. Parameters like length, width of processed area are the same, as well as Max height and Min height. It proves that the algorithm of the OptD method preserved the extreme values.

Decreased dataset is presented in Fig. 8.

Fig. 8. The fragment of TLS point cloud from one station for Biebrza river after OptD method application: a) top view, b) side view (source: own study in CloudCompare v.2.7.0).

At the first sight, there is no significant differences. They are visible in zoomed fragment (Fig. 9).

Fig. 9. The zoomed fragment of TLS point cloud for Biebrza river: a) original, b) after OptD method application (source: own study in CloudCompare v.2.7.0).

As it can be seen in Fig. 9 application of the OptD method resulted in more clear visualization, especially in those areas where there was very high density of measured points.

4 Conclusions

The OptD method works very well not only for ALS, but also for TLS. It is based on creation of strips (horizontal or vertical). Those strips are usually based on the measurement strips resulted from the way the object/area is scanned or could be determined by the object's shape, size, range etc. The strips can be analyzed in plane X_0Y (ALS and TLS, extensive object) or in Y_0Z or X_0Z (TLS, relatively small object). However, generalization algorithm works in each strip individually. The relative position of points within strip is analyzed in relation to Z coordinate (first case) or in relation to X or Y coordinate. In both variants the OptD method gives satisfying results.

On the basis of the presented studies showing the applications of the OptD method on TLS and ALS data, the following detailed conclusions were drawn:

- applying the OptD method for decreasing the datasets did not affected DTMs generated on the basis of these datasets – for ALS data coefficient of determination D is 0.98; for TLS data D is 0.99; differences in standard deviations SDs calculated for original and optimized datasets are 0.012m and 0.011m, respectively,
- difference in standard deviations SDs calculated for TLS data for extensive area is1.485m. It resulted from the fact, that it is was relatively flat area where more points was removed, therefore mean heights for original and reduced datasets are different.
- comparison of meshes generated from original and optimized dataset for relatively small objects do not show significant differences, they are visible when grid is generated,
- visualizations of original and optimized datasets shows differences when it is zoomed.

And the general conclusions are:

- The OptD method reduces the ALS and TLS data.
- Reduction of the dataset in many cases improves its readability, and consequently its visualization.
- Reduced dataset can be used to build DTM.
- The OptD method can be used by architects to inventory objects with complex structure.
- The OptD method can be used in CAPAP project.

References

1. K. Bakuła, Archiwum Fotogrametrii,Kartografii i Teledetekcji **23**, 19 (2012)
2. B. Bauer-Marschallinger, D. Sabel, W. Wagner, Computers & Geosciences **72**, 84 (2014)
3. L. Barazzetti, F. Remondino, M. Scoioni, *Proceedings of SPIE Optics+Photonics* **7447** (2009)
4. W. Błaszczak, *Optimization of large measurement results sets for building data base of special information system*. Doctors thesis (University of Warmia and Mazury in Olsztyn, 2006)
5. W. Błaszczak-Bąk, A. Janowski, W. Kamiński, J. Rapiński, Canadian Journal of Remote Sensing **37** **(6)**, 583 (2011)

6. W. Błaszczak-Bąk, A. Janowski, W. Kamiński, J. Rapiński, Journal of Indian Society of Remote Sensing **39**, 591 (2011b)
7. W. Błaszczak-Bąk, Acta Geodynamica et Geomaterialia **13/ 4(184)**, 381 (2016)
8. W. Błaszczak-Bąk, A. Sobieraj-Żłobińska, M. Kowalik, Meas. Sci. Technol. **28 (7),** 075009 (2017)
9. Y. Chen, *High performance computing for massive LiDAR data processing with optimized GPU parallel programming* (2012)
10. D. H. Douglas, T. K. Peucker, Canadian Cartographer **10(2)**, 112 (1973)
11. D. Gościewski, GIScience&Remote Sensing **50(5)**, 515 (2013)
12. E. Hadaś, A. Borkowski, J. Estornell, *The International Archives of the Photogrammetry, Remote Sensing and Spatial Information Sciences* **XLI-B8,** 629 (2016)
13. G. F. Jenks, Annals of the Association of American Geographers **17 (1)**, 1 (1981)
14. T. Kersten, K. Mechelke, L. Maziull, *The International Archives of the Photogrammetry, Remote Sensing and Spatial Information Science* **XL-5/W4**, 1 (2015)
15. H. Liu, X. Yang, *Software reuse in the emerging cloud computing era,* chapter 9 (Information Science Reference, 2012)
16. M. Pomykoł, M. Poniewiera, *Numeryczne projektowanie w geodezji górniczej* (Wyd. Politechniki Śląskiej, Gliwice, 2009)
17. F. Remondino, F. Menna, *The International Archieves of Photogrammetry, Remote Sensing and Spatial Information Science* **XXXVII (B5)**, 199 (Beijing, China, 2008)
18. M. Visvalingam, J. D. Whyatt, *Line generalization by repeated elimination of point* (Cartographic Information Systems Research Group, University of Hull, 1992)

E3S Web of Conferences **26**, 00006 (2018) https://doi.org/10.1051/e3sconf/20182600006
2017 BGC

The concept of estimation of elevator shaft control measurement results in the local 3D coordinate system

Daria Filipiak-Kowszyk[1], *Waldemar* Gierski[1], *Waldemar* Kaminski[1], and *Karolina* Makowska[1,*]

[1]Gdansk University of Technology, Faculty of Civil and Environmental Engineering, Narutowicza St. 11/12, Gdansk, Poland

Abstract. Geodetic control measurements play an important part because they provide information about the current state of repair of the construction, which has a direct impact on the safety assessment of its exploitation. Authors in this paper have focused on control measurements of the elevator shaft. The article discusses the problem of determining the deviation of elevator shaft walls from the vertical plane in the local 3D coordinate system. It presents a concept of estimation of measurements results base on the parametric method with conditions on parameters. The simulated measurement results were used to verify the concept presented in the paper.

1 Introduction

An important aspect in the context of using the structure is to ensure its safety. For this reason, geodetic control measurements are made in order to obtain information about the structure behaviour [1, 2].

The process of building an elevator shaft is a very demanding venture. The construction has to comply with a number of conditions to ensure its proper operation and safety in the future. The whole system of an elevator consist of a large number of devices and mechanical systems. Therefore, only their proper fitting in the shaft space and the optimum selection of dimensions allow for ensure the reliability and proper exploitation of the elevator shaft.

Design premises for this type of facility are subject to inspection at every stage of construction, particularly in the rough. Geodetic methods and instruments are important in the verification of elevator shaft geometry. Control measurements of elevator shafts are made to determine the deviation from the vertical plane and walls torsion of the structure [3, 4, 5]. The obtained results allow to specify the way of assembly of axes of the cabin and counterweight guide rails and to adjust the dimensions of the cabin in a way that allows for a free movement and fulfilment of the design premises [6, 7, 8].

In the traditional approach, observations are made in a local 2D coordinate system (XY) and only elements measured in relation to this coordinate system are adjusted. This article proposes the concept of estimation of an elevator shaft control measurements results in the local 3D coordinate system (XYH) where the conditions for adjustment at individual measurement stages are determined not only for two dimensions (XY) of the control points but also for elevations (H). This approach involves conditions of placement of control points in three dimensions, which allows for a complete set of information about the geometry of the elevator shaft walls.

The elevator shaft geodetic inspection is carried out on the basis of control base, which is installed on the foundation of the shaft, below the lowest stop level on which the cabine stops. The rectangular shape is usually four points [5, 6, 7, 8]. In practical applications these may also be two points. The scheme of the elevator shaft with the location of the points of the control base and control points are shown in Fig. 1. The measuring frame determines the plane XY , while the vertical axis of the assumed local coordinate system is realized by direction of action of gravity.

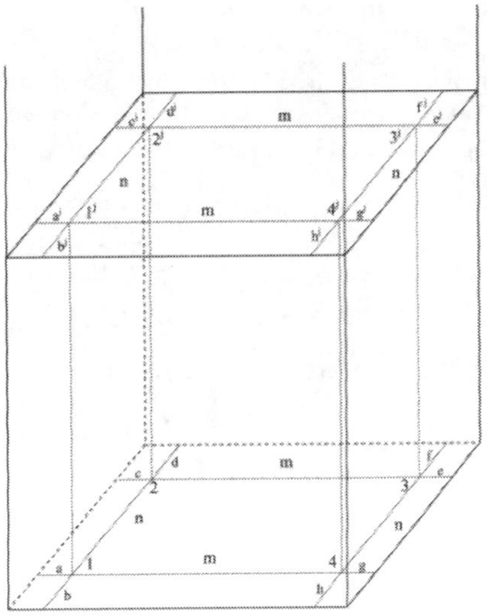

Fig. 1. Scheme of an elevator shaft.

* Corresponding author: karolina.makowska@pg.edu.pl

Determination of the deviation of the shaft walls from the vertical plane is based on the designation of adjusted XY coordinates of control points for each measuring level, while defining the deviation from the horizontal plane is based on the designation of the adjusted H coordinates of those points. For this purpose, the object of measurement are values on the reference level (a, b, c, d, e, f, g) and values on each measuring levels ($a^j, b^j, c^j, d^j, e^j, f^j, g^j$, where j – number of measuring level).

2 The theoretical basis

2.1 Parametric method with conditions on parameters

The method described in this article applies not only to the problem of determining the deviations of walls of an elevator shaft, but can also be use to solve other problems in engineering geodesy, what is shown in the publications [9, 10].

The most important information about this method, which will allow for understanding of the way to solve the problem, are presented below. Detailed analysis and considerations related to the parametric method with conditions on parameters can be found in [3, 4, 11, 12].

As in publications [3, 4, 9, 10], the problem can be expressed using the system of equations:

$$\begin{cases} \mathbf{AX} + \mathbf{L} = \mathbf{V} \\ \mathbf{BX} + \mathbf{\Omega} = \mathbf{0} \end{cases} \tag{1}$$

where:
$\mathbf{A}$, $\mathbf{B}$ - known coefficients' matrices,
$\hat{\mathbf{X}}$ - estimator of $\mathbf{X}$ vector,
$\mathbf{L}$, $\mathbf{\Omega}$ - free terms vectors,
$\hat{\mathbf{V}}$ - estimator of corrections' vector.

As shown in publications [11, 12], the solution of the above system of equations (1) is a stepwise solution. In the first stage, the vector of correlates $\hat{\mathbf{K}}$ is searched (2):

$$\hat{\mathbf{K}} = -\{\mathbf{B}(\mathbf{A}^T\mathbf{PA})^{-1}\mathbf{B}^T\}^{-1}\{\mathbf{\Omega} - \mathbf{B}(\mathbf{A}^T\mathbf{PA})^{-1}\mathbf{A}^T\mathbf{PL}\} \tag{2}$$

It allows to determine a vector of adjusted parameters $\hat{\mathbf{X}}$ (3):

$$\hat{\mathbf{X}} = -(\mathbf{A}^T\mathbf{PA})^{-1}(\mathbf{A}^T\mathbf{PL} - \mathbf{B}^T\hat{\mathbf{K}}) \tag{3}$$

which is necessary to obtain an estimator of corrections' vector $\hat{\mathbf{V}}$ (4):

$$\hat{\mathbf{V}} = \mathbf{A}\hat{\mathbf{X}} + \mathbf{L}. \tag{4}$$

According to previously quoted sources, the above solution may be replaced by a simplified solution. For this purpose, the equation $\mathbf{B}\hat{\mathbf{X}} + \mathbf{\Omega} = \mathbf{0}$ is treated as an additional equation of corrections, for which infinitely large values of weight $\mathbf{P}_\infty$ are assumed. From this there is a dependence (5):

$$\mathbf{B}\hat{\mathbf{X}} + \mathbf{\Omega} = \mathbf{V}_\mathbf{\Omega}. \tag{5}$$

This approach allows to obtain from the solution a correction value of zero. What is more, this allows to perform further calculations in a manner typical of the parametric method for which we can write a system of corrections in the form:

$$\tilde{\mathbf{V}} = \tilde{\mathbf{A}}\hat{\mathbf{X}} + \tilde{\mathbf{L}} \tag{6}$$

where:

$$\tilde{\mathbf{A}} = \begin{bmatrix} \mathbf{A} \\ \mathbf{B} \end{bmatrix} \quad \tilde{\mathbf{L}} = \begin{bmatrix} \mathbf{L} \\ \mathbf{\Omega} \end{bmatrix} \quad \tilde{\mathbf{V}} = \begin{bmatrix} \hat{\mathbf{V}} \\ \mathbf{V}_\mathbf{\Omega} \end{bmatrix}$$

The assumed weight matrix $\tilde{\mathbf{P}}$ is:

$$\tilde{\mathbf{P}} = \begin{bmatrix} \mathbf{P} & \mathbf{0} \\ \mathbf{0} & \mathbf{P}_\infty \end{bmatrix} \tag{7}$$

The simplified solution are presented below:

$$\hat{\mathbf{X}} = -(\tilde{\mathbf{A}}^T\tilde{\mathbf{P}}\tilde{\mathbf{A}})^{-1}\tilde{\mathbf{A}}^T\tilde{\mathbf{P}}\tilde{\mathbf{L}} \tag{8}$$

The use of the relation (8) allows the estimator of the correction's vector (6) to be written according to the relation:

$$\tilde{\mathbf{V}} = -\tilde{\mathbf{A}}(\tilde{\mathbf{A}}^T\tilde{\mathbf{P}}\tilde{\mathbf{A}})^{-1}\tilde{\mathbf{A}}^T\tilde{\mathbf{P}}\tilde{\mathbf{L}} + \tilde{\mathbf{L}} \tag{9}$$

2.2 Observation equations

For further analysis, the scheme of an elevator shaft shown in Fig. 1. was used.

In the case considered by the authors, the adjustment of acquired data is based on the use of parallelism and perpendicularity conditions that occur between the elements of an elevator shaft. The innovative character of the proposed method consists in linking the horizontal coordinates of control points to the heights of each level. Consequently, the estimation of the measurement results performed in the local 3D coordinate system using the parametric method with conditions on parameters allows simultaneous determination of corrections to the coordinates of all axes of the assumed system. This is important because other publications about the implementation of inventory measurement and adjustment of their results do not take into account the estimation of the next levels on which the measurement is made – heights are taken directly from the building project [6, 7].

Opposite walls of the elevator shaft should be parallel to each other, therefore, according to Fig. 1, the following parallelism conditions are assumed:

$$\begin{Bmatrix} \hat{X}_4 - \hat{X}_1 - k = 0 \\ \hat{X}_3 - \hat{X}_2 - k = 0 \end{Bmatrix}$$

$$\left.\begin{aligned}\hat{Y}_2 - \hat{Y}_1 - p = 0\\ \hat{Y}_3 - \hat{Y}_4 - p = 0\end{aligned}\right\}$$

where:
$\hat{X}_1, \hat{Y}_1, \hat{X}_2, \hat{Y}_2, \hat{X}_3, \hat{Y}_3, \hat{X}_4, \hat{Y}_4$ - adjusted coordinates of control points.

What is more, the condition of perpendicularity between individual walls should also be maintained:

$$\left.\begin{aligned}\hat{X}_2 - \hat{X}_1 = 0\\ \hat{X}_3 - \hat{X}_4 = 0\end{aligned}\right\}$$

$$\left.\begin{aligned}\hat{Y}_4 - \hat{Y}_1 = 0\\ \hat{Y}_3 - \hat{Y}_2 = 0\end{aligned}\right\}$$

Observation equations can be written according to the following relations:

$$\hat{X}_1^j = a - a^j \qquad \hat{X}_2^j = c - c^j$$

$$\hat{X}_3^j = c + m + e^j \qquad \hat{X}_4^j = a + m + g^j$$

$$\hat{Y}_1^j = b - b^j \qquad \hat{Y}_2^j = b + n + d^j$$

$$\hat{Y}_3^j = h + n + f^j \qquad \hat{Y}_4^j = h - h^j$$

In this paper, in addition to the X, Y coordinates the heights H was also analyzed. Due to the contemplated assumption of a constant distance between the various measurement levels, the condition imposed on height is (10):

$$\hat{H}_i - js = 0 \qquad (10)$$

where:
i – number of control point,
j - number of measurement level,
s - assumed distance between measurement levels.

3 An example of practical use

To verify the presented concept, simulated data was used.

The considerations were divided into two variants:
- variant 1: A base of four points has been adopted and $H = 0$ for reference level;
- variant 2: A base of two points has been adopted and $H = 0$ for reference level.

In addition, the following conditions were applied for the variants:
- for first measurement level $\hat{H}_i - 2800 = 0$,
- for second measurement level $\hat{H}_i - 5600 = 0$,
- for third measurement level $\hat{H}_i - 8400 = 0$,
- for fourth measurement level $\hat{H}_i - 11200 = 0$.

3.1 Variant 1

In variant 1, four points of base, designated from 1 to 4, were used for analysis. Calculations were made at the reference level and four measurement levels j ($j = 1, 2, 3, 4$). It has also been assumed that the distance between the individual measurement levels should be 2800 mm.

The analyzes were based on the values shown in Table 1 and:

$m = 1200$ mm, $n = 1400$ mm,

$k = 1900$ mm, $p = 2000$ mm.

Table 1. Simulated data adopted to analysis – variant 1 ([mm]).

Measurement level	Number of base point							
	1		2		3		4	
	a	*b*	*c*	*d*	*e*	*f*	*g*	*h*
0	350	300	350	300	350	300	350	300
1	370	286	340	290	366	309	336	290
2	354	322	355	324	334	281	367	284
3	310	320	378	285	340	314	352	310
4	314	324	367	294	342	310	358	320

Adjusted coordinates of the control points, deviations from the vertical of the shaft walls and deviations from the reference level derived from the solution described above are summarized in the Table 2, Table 3 and Table 4 (relation (9)).

Table 2. Adjusted coordinates of the control points.

Number of point	Coordinate					
	$\hat{X}$ [mm]	$\hat{Y}$ [mm]	$\hat{H}$ [mm]			
			$j=1$	$j=2$	$j=3$	$j=4$
1	0.4	-3.1	2800.0	5600.0	8400.0	11200.0
2	0.4	1996.9	2800.0	5600.0	8400.0	11200.0
3	1900.4	1996.9	2800.0	5600.0	8400.0	11200.0
4	1900.4	-3.1	2800.0	5600.0	8400.0	11200.0

Table 3. Deviations from the vertical plane of the shaft walls ([mm]).

$\hat{\mathbf{V}}$	Measurement level			
	1	2	3	4
$d\hat{X}_1$	20.4	4.4	-39.6	-35.6
$d\hat{Y}_1$	-17.1	18.9	16.9	20.9
$d\hat{X}_2$	-9.6	5.4	28.4	17.4
$d\hat{Y}_2$	6.9	-27.1	11.9	2.9
$d\hat{X}_3$	-15.6	16.4	10.4	8.4
$d\hat{Y}_3$	-12.1	15.9	-17.1	-13.1
$d\hat{X}_4$	14.4	-16.6	-1.6	-7.6
$d\hat{Y}_4$	-13.1	-19.1	6.9	16.9

Table 4. Deviations from the reference level ([mm]).

$\hat{\mathbf{V}}$	Measurement level			
	1	2	3	4
$d\hat{H}_1$	-10.0	-12.0	-4.0	-16.0
$d\hat{H}_2$	-2.0	-8.0	-11.0	-8.0
$d\hat{H}_3$	-14.0	-18.0	-9.0	-9.0
$d\hat{H}_4$	-6.0	-3.0	-15.0	-22.0

3.2 Variant 2

The analysis was based on a base of two points: 1 and 3 (according to the Fig. 1). As in variant 1, four measurement levels $j = 1, 2, 3, 4$ and a distance of 2800 mm between them were assumed. For the analysis, the same simulated data of measured values were accepted, because of the willingness to compare the obtained results from both variants (Table 5).

Table 5. Simulated data adopted to analysis – variant 2 ([mm]).

Measurement level	Number of base point							
	1				3			
	a	$m+g$	b	$n+d$	e	$m+c$	f	$n+h$
0	350	1550	300	1700	350	1550	300	1700
1	370	1536	286	1690	366	1540	309	1690
2	354	1567	322	1724	334	1555	281	1684
3	310	1552	320	1685	340	1578	314	1710
4	314	1558	324	1694	342	1567	310	1720

The results of the calculations are shown in Table 6, Table 7 and Table 8.

Table 6. Adjusted coordinates of the control points.

Number of point	Coordinate					
	$\hat{X}$ [mm]	$\hat{Y}$ [mm]	$\hat{H}$ [mm]			
			$j=1$	$j=2$	$j=3$	$j=4$
1	4.3	-4.8	2800.0	5600.0	8400.0	11200.0
3	1904.2	1995.3	2800.0	5600.0	8400.0	11200.0

Table 7. Deviations from the vertical plane of the shaft walls ([mm]).

	Measurement level			
	1	2	3	4
$d\hat{X}_1$	24.3	8.3	-35.7	-31.7
$d\hat{Y}_1$	-18.8	17.2	15.2	19.2
$d\hat{X}_3$	-11.8	20.2	14.2	12.2
dY_3	-13.7	14.3	-18.7	-14.7

Table 8. Deviations from the reference level ([mm]).

	Measurement level			
	1	2	3	4
$d\hat{H}_1$	-10.0	-12.0	-4.0	-16.0
dH_3	-14.0	-18.0	-9.0	-9.0

4 Summary

The method of adjustment of observations made in the local 3D coordinate system presented in the article allows to perform more detailed analyzes of the geometry of the elevator shaft. The use of appropriate geometric conditions enabled the determination of such coordinates of adjusted control points for which the deviation of the shaft walls from the vertical plane obtained at the subsequent measurement levels was as small as possible.

It should be note that for uniform observational data, the final values of the adjusted coordinates and the deviations of the wall from the vertical plane are different – differences up to 4 mm. This value allows both methods to be considered correct, but for variant 2 the number of necessary stations of the instrument is smaller, which is an undoubted advantage of this method.

The calculations presented in the article were performed on simulated data, but it can be assumed that the proposed method can be applied successfully to determine the deviations from the vertical plane of the elevator shaft walls also for the data obtained for a real structure.

References

1. W. F. Caspary, *Concepts of network and deformation analysis* (Monograph 11, School of Geomatic Engineering, The University of New South Wales, UNSW Sydney NSW 2052 Australia, 2000)
2. I. M. Abdullahi, N. A. Yelwa, International Journal of Science and Technology **6 (1)**, 4 (2016)
3. M. Grala, *Optymalizacja wpasowania osi prowadnic dźwigowych w szybach windowych* (Konferencja Zakładów i Katedr Geodezji na Wydziałach Niegeodezyjnych, 1983) (in Polish)
4. M. Grala, G. Kopiejewski, A. Wasilewski, *Geodezja Inżynieryjna. Działy wybrane.* (Skrypty Akademii Rolniczo-technicznej w Olsztynie, Wydawnictwo ART, Olsztyn, 1986) (in Polish)
5. CH. Schittich, *In detail designing circulation areas: stairs, ramps, lifts: routing, planning principles* (Institut für internationale Architektur- Dokumentation GmbH & Col. KG, Munich, 2013)
6. M. S. Nesterionok, V. N. Veksin, Nauka i Tehnika **0 (2),** 38 (2015)
7. A. Piątkiewicz, H. Urbanowicz, *Dźwigi elektryczne,* (WNT, Warszawa, 1966) (in Polish)
8. *PN-EN 81-20:2014-10, Przepisy bezpieczeństwa dotyczące budowy i instalowania dźwigów -- Dźwigi przeznaczone do transportu osób i towarów -- Część 20: Dźwigi osobowe i dźwigi towarowo-osobowe*
9. D. Filipiak, W. Kamiński, Reports on Geodesy and Geoinformatics **97 /2014**, 71 (2014)
10. W. Kamiński, Reports on Geodesy **94/2013**, 47 (2013)
11. L. W. Baran, *Teoretyczne podstawy opracowania wyników pomiarów geodezyjnych* (Wydawnictwo Naukowe PWN, Warszawa, 1999) (in Polish)
12. Z. Wiśniewski, *Rachunek wyrównawczy w geodezji (z przykładami)* (Wydawnictwo UWM, Olsztyn, 2005) (in Polish)

https://doi.org/10.1051/e3sconf/20182600007

Geodetic survey as a means of improving fast MASW (Multichannel Analysis Of Surface Waves) profiling in difficult terrain/land conditions

Rafał Matuła[1,*], *Paulina* Lewińska[2]

[1] AGH University of Science and Technology, Faculty of Drilling, Oil and Gas, Krakow, Poland
[2] AGH University of Science and Technology, Faculty of Mining Surveying and Environmental Engineering, Krakow, Poland

Abstract. This paper revolves around newly designed and constructed system that can make 2D seismic measurement in natural, subsoil conditions and role of land survey in obtaining accurate results and linking them to 3D surface maps. A new type of land streamer, designed for shallow subsurface exploration is described in this paper. In land seismic data acquisition methods a vehicle tows a line of seismic cable, lying on construction called streamer. The measurements of points and shots are taken while the line is stationary, arbitrary placed on seismic profile. Exposed land streamer consists of 24 innovatory gimballed 10 Hz geophones. It eliminates the need for hand 'planting' of geophones, reducing time and costs. With the use of current survey techniques all data obtained with this instrument are being transferred in to 2D and 3D maps. This process is becoming more automatic.

1 Introduction

Since the CMP-style (Common Middle Point) methodology was introduced into engineering seismic profiling, improving the efficiency of seismic shallow data acquisition was discussed. It has been an important factor for most seismic production and research projects. Improvements can be discussed in regard to: automation, mobility, consistency and safety of seismic sources. Advancements in engineering technologies facilitated all of those factors to some extent. Over years, shallow seismic acquisition methods, and for the most part acquisition equipment, have been reduced in size and shape so to resemble prospecting equipment and methods. This has mainly been a necessity for the economy of small scale projects. Recent developments in shallow seismic imaging efficiency have taken advantage of high-resolution, surveying methods. This includes automated systems which incorporate sources and receivers designed for land applications with towed techniques, adopted from marine seismic acquisition. With combining modern, engineering acquisition systems' components and an awareness of the shallow seismic technology's potential, the acquisition costs (measured by required time) could be reduced up to 50 percent. Several experimental land acquisition systems have been developed to atomise following processes: roll and unroll seismic cable, plant and retrieve conventional spiked geophones into ground, haul fixed spreads of receivers along the ground surface maintaining source and receiver offset without invasive, manually positioned geophones. Researchers for over two decades attempted to find a way to tow long receiver spreads across the ground surface in a way similar to marine sea-streamers. This would allow to transfer efficiency of seismic surveying to land [2], [3], [8] Shear wave land streamers were also developed and tested for near-surface applications during the late 1990s [9]. Land streamers, that towed arrays or spreads, have primarily been designed and tested for applications of so called continuous seismic profiling. Low signal-to-noise ratios and the need for achieving highest frequency signals possible, are characteristic to shallow seismic surveys. Their volume has limited the widespread use in many near-surface settings of land streamer technologies. It is worth to mention that towed spread designs are suited for the high signal-to-noise ratio and lower frequencies.

This paper describes a towed spread developed for multichannel analysis of surface waves (MASW) [1]. The spread is designed to be hauled by the source vehicle and pressure coupled to various surfaces (varying from asphalt and lightly vegetated to loose soils). Comparisons of data acquired with the towed spread and conventional, coupled geophones is also given in this article. Appraise between the trade-off conventional "planting" of geophones for surface wave registration is described.

Nowadays more emphasis is being put on combing data from differed sources in to one 2D or 3D model. Specialists from various technical fields see the need of interdisciplinary works that would create one, complete picture of surveyed objects and structures with regard to there special placement. This is why geodetic/survey personnel is often being welcomed into data acquisition groups – since it can provide a space for correlation and comparison of various datasets. Described new data acquisition method is no different at this point. In order to

*Corresponding author: rmatula@agh.edu.pl

https://doi.org/10.1051/e3sconf/20182600007

connect the subsurface analysis performed by geophysics to pre-existing topographic maps GPS [4], [5] survey was required. It had to be done within time frame, and conditions facilitated by this new and fast geophysics method. What is more, all gathered information had to be implemented and stored in to one special data file thus allowing for fast access to all data. GPS receivers allow fast acquisition of spatial data curtail for correctly placing obtained geological profiles.

2 Device description

The prototype of described land streamer system is built with parts intended for application in various, conditions and acquisition techniques. Self-orienting, gimbal-mounted geophones with 10 Hz sensors on half spherical underlays were recording the seismic wave arrivals. The receivers were mounted in ~2.5 kg underlays to ensure geophone-to-ground coupling. The mentioned geophones with underlays were connected by belts made from synthetic material, resistant to stretching. Signals were transferred to seismograph by cable that had takeout's for 48 geophones at 2 m intervals. Cables are terminated with robust waterproof connectors. The towed land streamer was limited by the 48-channel capacity of these connectors. The projected streamer construction provides possibility of adding an extra weight in case a need of increasing geophone-to ground coupling appears. Performed experiments have demonstrated that geophone-to-geophone coupling via the half-sphere underlays is possible and very effective for the porpoise of surveying open, natural areas. To take full advantage of the efficiency of the towed land streamer acquisition system, a fast, self-propelled sources that provide strong signal (characterized by high and low frequencies) was required [19].

For surveys described in this article a sledge-hammer based, Seismic Source Company source was used. Mentioned seismic source was mounted on tracked tractor (IHI35), which was used for towing land-streamer. Used acquisition system was configured for imaging relatively shallow and intermediate depth targets, simultaneously. Half-automatic scanning of the subsurface using surface waves can be done cost effectively and with confidence at relatively high production rates if: consistent time-lapse measurement can be made, high production rates can be maintained, and fixed processing flows can be used on all data sets from a particular area.

Fig. 1. In section A – part of projected land streamer, B – seismic source used for surveys.

Acquisition test have proven that coupling necessary for accurate recording of surface waves can be established and maintained over an extended distance and variable terrain with just pressure contact to the earth's surface through aggressive contact points. Unlike standard surveying, where coupling of geophones by direct "planting" produces the processed data, surface wave energy recorded from planted geophones is equivalent (for most applications) to surface wave energy recorded from pressure coupled geophones.

3 Brief description of device application procedure

Acquisition schema with the towed land streamer is similar to technique on which the marine surveys was primarily based [3, 4, 6, 9]. In contrast to the continuous movement of marine survey systems, the operational procedure of the towed land streamer is based on a "excitation technique at a standstill" tactic with some projected acquisition step [17, 7, 8, 14, 15]. In Figure 1A, the streamer comprises of one module, with forty eight 2 m takeout's is shown. Initially in standard survey methodology case study, layout of receivers is kept at a fixed location, and shot positions are moved from back to front of the seismic spread (Fig. 2B) [5, 10]. In comparison to land streamer, once the shot position reaches the front, it is kept at a fixed distance related to the first receiver's position. Sources and receivers are shifted simultaneously after each shot (Fig. 2A) [1, 2]. The streamer is moved up one shot interval, the shot is detonated once the streamer has been stationary for a few seconds, and the process repeated up to the end of profile. Since moving the land streamer generates significant noise, including variable vertical accelerations, there is no possibility to shoot while it is in motion. Comparable acquisition procedures using spike-mounted geophones and standard common-midpoint (CMP) cables are more labour intensive [10, 13, 18]. To record with 48 active channels (as for the exposed land-streamer example), a minimum of two 24 takeout engineering cables is required. At least one 48- channel standard cable is necessary for conventional procedures using a roll-along device to change active channels [16].

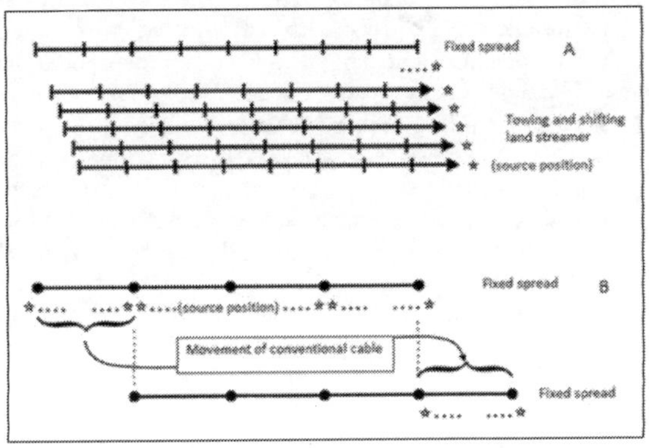

Fig. 2. Land streamer measurement realization (A), CMP-style surveying (B).

One profile was undergoing detailed examination. The survey geometry was modified for obtaining very high-resolution data. 2 m receiver spacing was used for every acquisition schema. Each shot-points were separate in every 12 meters for roll-along technique, with seismic spread movement equal to ¼ spread length (schematic example in Fig. 2B). Land streamer survey interval was at 5 meter between each position of projected acquisition system. Measurements were made for surveys with the sledgehammer-type source described in the section above.

In general, line preparation and deployment and the retrieval of geophones and cables are most time consuming aspects of conventional strategy of seismic surveys. For an efficient conventional shallow seismic survey on land, a field crew of five people is needed: one person operating the recording equipment, two managing the seismic source, two geophone/cable layers. For a comparable land-streamer survey, only three members of field personnel are required (as during the traditional survey) but as operator doubles since the driver of the towing vehicle and the geophone/cable layers are not required. Compared to the traditional approach, the towed land-streamer system was about 50–90% faster and required only about 30–40% need of personnel. In other words, the land streamer surveys on the same profile were completed with 25–50% of the person-hours required for the traditional surveys. During described test about 200 records were obtained during four hour land streamer measurements. Using the half-automatic sledge hammer source, acquisition speed of the towed land-streamer seems to be faster (~2 times without repeating) than the vibrator technique.

4 Comparison of towed spread geophones and conventional spike coupled geophones

Evaluating how effective a towed spread records seismic data is relative to conventionally planted geophones can be done through analysis of comparable data. Ideally these data are recorded using the same source energy and as nearly consistent ground surface conditions as possible. For this study, two records – part of CMP-style record and landstreamer- made record, were selected with equal receiver positions and uniquely near-surface conditions and characteristics of part of seismic profile. Both data were simultaneously recorded using both spike planted geophones and pressure coupled streamer geophones. For the Polish lowland site both receiver configurations were coupled to as near identical soil as possible. However, under natural condition conventional spread was planted above the road track while the towed spread was drug along road's surface. Underground structure of chosen site has relatively small differences, and can be used as test site for planted and streamer geophones. It is almost completely without near-surface static inconsistency between adjacent receiver locations. Surface wave analysis usually involves calculation of dispersion. It is the first and most efficient diagnostic step in the processing flow of surface wave data [3]. Subtle wavelet differences are evident in waveforms and arrival consistency when comparing equivalent shot gathers from planted geophones and geophones in the streamer. Due to the reduced trace-to-trace consistency in waveforms on the streamer data, the amplitudes appear to vary more than spectral analysis suggests from dispersion energy analysis [3]. Seismic data designed and acquired for surface wave analysis were recorded using the same seismic source energy for both the towed array and conventional spike-planted geophone spreads.

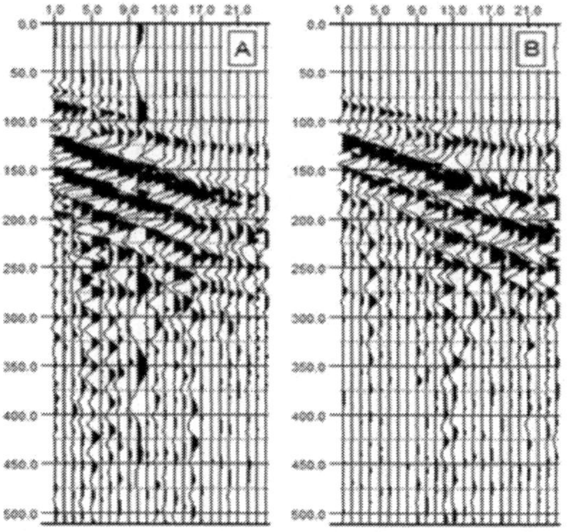

Fig. 3. Examples parts (half of all active channels of central part of records) of obtained records. In section A data recorded by using land streamer. In section B, normal surveying with spiked into ground geophones. On the vertical axis time in [msec].

Geophones recording surface waves on the road surface using the towed array have a slightly lower dominant frequency, higher ratio of surface waves to body waves, and a much more coherent and fully developed dispersive surface wave. It is unlikely the coupling is improved with the land streamer, but recording on the road surface does improve the quality (trace-to-trace consistency and amplitude) and quantity of surface wave energy captured. Dispersion energy analysis reinforces this observation with a much larger amplitude and better defined fundamental mode than with the land streamer data. An apparent break in slope of the surface wave arrival near the middle of the spread appears to manifest itself on dispersion energy analysis as the break in slope observed at about 15 Hz. Increased noise at higher frequencies on conventional geophones is evident in dispersion energy analysis. Clearly the road represents much more uniform media than the variably sculpted, eroded, and infilled road rut, therefore producing a dispersion curve with minimal interference from trace-to-trace shift.

 https://doi.org/10.1051/e3sconf/20182600007

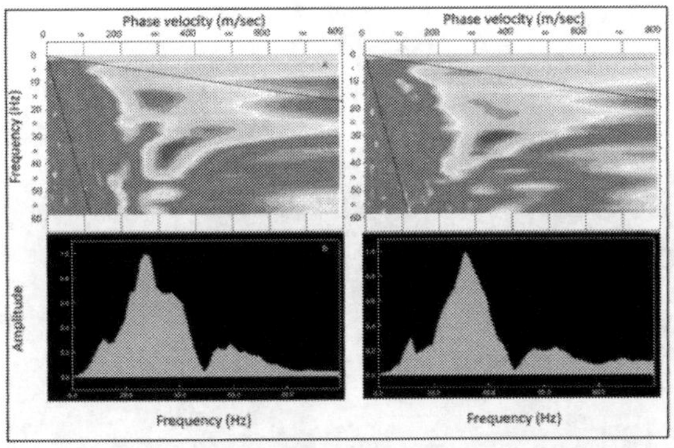

Fig. 4. A and B spectrum and dispersion curve of spiked geophones data. C and D the distribution for land streamer data.

During presented tests a RTK-GPS survey of measured lines was performed. The technique employed in both methods was the same however during MASW survey it needed to be much faster and done almost on the go. GPS survey consisted of placing 3 GPS reference stations that would cover the area of geophysics works. This allowed the crew to be independent from Polish GPS corrections sourced during the actual surveys. It was checked beforehand that the area was placed unfavourably in regard to them. The main task was to cared out GPS-RTK surveys during the movement of geophone-units every few meters in order to establish position of the line and selected units.

Also it was important not to get in the way with geophysics crew. GPS data was downloaded on the side and implemented in to further processing's. All gathered data, survey and geophysics, ware connected in to one 3D model and stored as one data set. Also georeferenced 2D maps an cross-sections ware created. This allows for better visualisation on how subsoil structure influences the surface and constructions placed upon it.

5 Conclusions

In situations where Multichannel Surface Wave Analysis (MASW) is used as a main methodology for image of subsurface material, a towed array on the paved surface can produce better data than spike-planted geophones in natural terrain conditions. For surface wave analysis where the ground surface is conducive to spike-planted geophones, a towed spread produces equivalent data quality. All aspects considered, this towed spread design is very effective in recording good consistent surface wave energy in most surface settings. With the source as the tow vehicle and the seismograph attached to the rear of the spread, data can be acquired rapidly and surface wave processing can be accomplished using fixed flows in almost real time. Body wave analysis will in many surface settings be possible, providing good preliminary images and maps of the surface and adding GPS coordinates will allow for better 3D placement of acquired data. What is more coordinates can be downloaded on side and be implemented during post-processing and visualisation of geophysics data thus allowing other specialist for better interpretation of the results.

The project was funded by statutory funds of AGH no. 11.11.150.005 and 11.11.190.555.

References

1. A.J. Kruppenbach, J.W. Bedenbender, *Towed land cable: U.S. Patent 3 923 121* (1975); *Towed land cable: U.S. Patent 3 954 154* (1976)
2. A.J.M. Pugin, T.H. Larson, S.L. Sargent, J.H. McBride, C. E. Bexfield, The Leading Edge **23 (7)**, 677 (2004)
3. C.B. Park, J. Xia, R.D. Miller, *68th Ann. Internat. Mtg. Soc. Expl. Geophys., Expanded abstracts,* 1377 (1998)
4. D. Einarsson, L. Brooks, G. Bennett, A. White, Geophysics **42**, 148 (1977) – abstract
5. D.W. Steeples, Mining Engineering, 1297 (October 1993)
6. I. Gausland, The Leading Edge **19 (8)**, 903 (2000)
7. J. Ivanov, R.D. Miller, P. Lacombe, C.D. Johnson, Geophysics **71**, A39 (2006)
8. J.M. Lorenzo Saanumi, C. Westbrook, S. Egnew, *76th Ann. Internat. Mtg. Soc. Expl. Geophys., Expanded abstracts,* 1496 (2006)
9. J. Caldwell, W. Dragoset, The Leading Edge **19 (8)**, 898 (2000)
10. K.H. Waters, *Reflection seismology – A tool for energy resource exploration* (New York, 1987)
11. K. Maciuk, Geodetski vestnik **60 (3)**, 455 (2016)
12. K. Maciuk, Artificial Satellites **51 (1)**, 1 (2016)
13. M.A. Speece, S.J. Betterly, R.H. Levy, *76th Ann. Internat. Mtg. Soc. Expl. Geophys., Expanded abstracts,* 1 (2007)
14. M. Van der Veen, A.G. Green, Geophysics **63**, 1408 (1998)
15. M. Van der Veen, Buness, H.A. Buker, A.G. Green, J. Env. Eng. Geophys. **5**, 39 (2000)
16. T. Ghunaim Al-Anezi, M. AlMalki, T. Alkhalifa, Arabian Journal of Geosciences **6 (1)**, 235 (2013)
17. T. Inazaki, *Ann. Symp. Environ. Engin. Geophys. Soc. (SAGEEP), Expanded abstracts,* 207 (1999)
18. T. Inazaki, *Near Surface 2006, Expanded abstracts,* A016 (2006)
19. Y. Sawasdeen, Geophysical Prospecting **57 (1)**, 125 (2009)

https://doi.org/10.1051/e3sconf/20182600008

Integration of thermal digital 3D model and a MASW (Multichannel Analysis of Surface Wave) as a means of improving monitoring of spoil tip stability

Paulina Lewińska[1,*], *Rafał* Matuła[2], *Artur* Dyczko[3]

[1] AGH University of Science and Technology, Faculty of Mining Surveying and Environmental Engineering, Krakow, Poland
[2] AGH University of Science and Technology, Faculty of Drilling and Petroleum Engineering, Krakow, Poland
[3] The Mineral and Energy Economy Research Institute of the Polish Academy of Sciences, Krakow, Poland

Abstract. Spoil tips are anthropogenic terrain structures built of leftover (coal) mining materials. They consist mostly of slate and sandstone or mudstone but also include coal and highly explosive coal dust. Coal soil tip fires cause an irreversible degradation to the environment. Government organizations notice the potential problem of spoil tip hazard and are looking for ways of fast monitoring of their temperature and inside structure. In order to test new monitoring methods an experimental was performed in the area of spoil tip of Lubelski Węgiel „Bogdanka" S.A. A survey consisted of creating a 3D discreet thermal model. This was done in order to look for potential fire areas. MASW (Multichannel analysis of surface wave) was done in order to find potential voids within the body of a tip. Existing data was digitalized and a 3D model of object's outside and inside was produced. This article provides results of this survey and informs about advantages of such an approach.

1 Introduction

Spoil tips are anthropogenic terrain structures built of unusable rock material excavated during underground or open pit mining. Usually the term spoil tips refer to coal spoil tip. As such they consist mostly of slate and various types of sandstone or mudstone, also include coal and coal dust. Depending on the percentage amount of mentioned elements spoil tips can are susceptible to fires [5], [7].

Coal tips fires could be caused externally or internally. External cause is relatively easy to eliminate – usually during construction and revegetation those structures are closed for public access or open only to some extent. However there are still a few that belonged to closed mines and could be viewed as a constant treat. Most tip fires begin inside the structure. If a stack effect is introduced within the body of the tip air movement can induce coal dust to explode. If enough coal is located in the vicinity of the explosion it can caught on fire. Stack effect occurs when there are holes in the tip, usually vertical, chimney-like, that allowed for air to move from the outside thru spoil tip with large speed. This also provides oxygen for the fire that started inside. Depending on the size of the tip, amount and kind (age, fraction, placement etc.) of flammable material and amount of oxygen being transferred inside the speed in with the fire grows varies. It can take up to few years before signs of fires become visible as flames or smoke. In a meantime tip's stability degradation, contamination of the water etc. can occur. Stability can be so demolished that fire brigades are unable to enter the structures safely while trying to extinguish the source. Since there are usually a few sources of fire hidden inside and access to them limited stopping the fire may take years rather than weeks or days. Thus prolonging the degradation of the environment [4].

Spoil tips build currently are constructed in a way that should mitigate or a least minimalize possibility of fires. It is worth to mention that during past few years coal excavation in Poland has changed, becoming more and more effective. Nowadays lower coal seams are being excavated and there is much less leftover material containing coal. Also more emphasis is put on localisation of spoil tips. They are being constructed, if possible, further from human habitats and lakes, rivers, springs etc. The latter being important due to the fact that internal fires ware often fumed with oxygen coming from vaporising water. Rain water is being transported from the tip. Also way of stacking is planned and monitored. Machines compress the material so to avoid creating voids and waste material is mixed to minimise coal concentration.

Currently some spoil tips are undergoing monitoring process. However this is usually done in a traditional way. Responsible personnel walks the area and looks for fire, land slide or holes in the structure. In winter this survey also includes looking for places with vast vegetation or lack of snow cover. This may be unsafe for person responsible since spoil tips burn from the inside usually seem normal on the outside. Stepping on a fine cover may cause the hole to appear or poisonous gas to emerge [7,15,17].

In the view of recent spoil tips fires and landslides of tips within open pit mines some government organizations

* Corresponding author: lewinska.paulina@gmail.com

 https://doi.org/10.1051/e3sconf/20182600008

and mines see the potential in creating systems for constant or periodic monitoring of their geometric and thermal stability. This article provides a combined survey, thermal and geophysics method of monitoring the tip that could be implemented within mine's environmental hazards monitoring system.

2 Spoil tip of LW Bogdanka coal mine

In order to test new ways of observing coal spoil tips an experimental monitoring technique was performed in the area of spoil tip of Lubelski Węgiel „Bogdanka" S.A. hard coal mine in December of 2015. LW "Bogdanka" hard coal mine is located in the east part of Poland. Its coal deposit stretches alongside polish border and crosses to Ukraine. Bogdanka is one of the youngest polish coal mines, only underground hard coal mine and only mine excavating this deposit on polish side of the border.

Spoil tip of "Bogdanka" coal mine is supplied with two kinds of waste material. One comes from excavation process (waste rock longwall excavation method) and drilling new corridors, other from coal handling and preparation plant. In order to facilitate multiple local law regulations (national park, local govern plans etc.) during recent years a lot of work has been put into using as much waste material as possible in eco-friendly methods (landfill material, clinker brick factory etc.). This is why while an increase in waste material production has been noticed the spoil tip growth was within plans. Stored material can be devised as fallows 35 % of claystone of non or minimum cleavage with little organic matter, 20 % of dark grey shale and claystone with clear cleavage with higher content of organic matter, 20 % of grey mudstone with no cleavage, 10 % of siderite 15 % of mudstone and sandstone. Average grainsize: 40–50 % of coarse gravel (200÷20 mm), 30–40 % of sand (20–0.5 mm) and slit (less than 0.5 mm) [10].

The object covers the area of 88.56 ha, however only 76 ha is the actual tip. The rest is being used as emergency roads, drainage dich etc. The current height is between 16 and 29 metres. Planed height could be up to 90 metres. Currently most of the bulking is being performed in the east part of the tip. For the past 12 do 20 years South-West slopes have been planted with various threes in order to ensure stability [10].

3 3D and thermal survey of Bogdanka spoil tip

In order to perform a proper diagnostic of a coal spoil tip four factors need to be taken in to account: slope angle and stability, temperature on the surface of the tip (possible anomalies), interior structures (existence/lack of voids a holes). Obtaining necessary information and taking in to account tips geological structure, water conditions and its history allows for not only establishing the current state of the structure but also predicting possible dangerous changes in its stability.

During this study three surveys were performed simultaneously.

3.1 3D discreet model – Digital Terrain Model

First step of the project was to create a base – a model of the structure – that all other (thermal, geometry, geophysics) data could be tied or implemented into. In this case it was necessary to create a Digital Terrain Model (DTM) – 3D representation of a terrain's surface defined as a discreet representation of terrains height and includes and interpolation algorithm that would allow to recreate its shape [2]. In this case survey had to be done during bulking process and in the way least disturbing for maintains crew.

Fig. 1. Point cloud of "Bogdanka" spoil tip done with RIEGL laser scanner [10].

It was decided that the fastest way of generating a discreet model of such an object would be to use laser scanner [8]. Laser scanners are devices that allow for measuring distance and angle to numerous points of an object and estimating there placement in the local coordinates (registered) system. Later, laser scans done from separate scan station could be connected with the use of targets – natural or pre-prepared elements of known geometry. If coordinates of those elements are measured (with GPS etc.) point cloud could be transformed into global coordinates system [11].

In this case it was decided that a long range scanner should be tested [10]. This allowed the survey crew to enter the measured, potentially dangerous object only in a limited way. Five scans were performed with the use of RIEGL laser scanner (range up to 10 km). Scans were registered automatically with the use of natural elements. This also allowed for limiting access of the crew to the object (Fig. 1).

Those scans have become the base for all further survey and can be viewed as discreet Digital Terrain Model. They covered the general shape of the structure but did not give detail information on every slope. Spoil tips structure is complicated with multiple higher and lower points and various drainage diches. Also some of the transporting machines provided an obstacle for the scanner. Nevertheless this scan season provided an excellent reference model for further analysis.

 https://doi.org/10.1051/e3sconf/20182600008

3.2 Temperature of a spoil tip – Thermal Digital Terrain Model

Thermal Digital Terrain Model is a type of DTM with added information on surface temperature. In case of discreet DTM it simply means that an extra column, representing temperature is added.

There are multiple ways of obtaining value of temperature of an object. In this case a remote method had to be used due to the size of the object, time management and minimising crew presence on the tip. Remote ways include infrared cameras that detect and register the radiation in the long-infrared range of the electromagnetic spectrum. Then Stefan-Boltzmann or Planks laws that bound the heat emitted from the object with object's temperature are being used. Cameras used in thermography measure strength of emission of electromagnetic waves (that come from a certain direction). The value of temperature depends on the strength of received signal. This is called radiation flux. All objects emit a certain electromagnetic strength of long-infrared wave [16,18].

It is possible to create a Thermal DTM with the use of pictures an scans done from a plane, using standard photogrammetric method [1], [2], [3]. However the aim of this study was to create a system that could be used in all conditions and did not require any outside help. This is why a new Z+F IMAGER 5010 laser scanner equipped with T-Cam Rev 1.0 thermal camera was used. This camera is calibrated with laser scanner and provides the same geometric results as internal camera. It is possible to cover the point cloud with colours from thermal images automatically provided the usage of appropriate Z+F software. This point cloud can be later exported in to other formats and edited [9], [10].

This laser scan was performed on two, potentially most susceptible to fires and lads slides parts of the slopes. The result was 26 laser scans registered manual with the use of natural elements. Later on those scans were registered again and connected to the point cloud from RIEGL laser scanner so to create a proper 3D thermal model of the structure.

Results of such an experiment were satisfactory to some extent. There was little problem with registering the long range scans, they connected not only with the use of the elements of the spoil tip but also nearby buildings and mines chimneys were used. However registering Z+F scans has proven to be difficult due to the lack of easily identifiable elements in the vicinity. Also the range of the T-Cam Rev 1.0 camera was not satisfactory at all times. Scan stations in some cases had to be place in 10 to 20 metre distance from the edge of the spoil tip. Distance from the scanner to the top was more than 50 metres which was apparently to fair for cameras software. In most scans some thermal data was cut off thus living top parts of the tip without thermal information (point cloud existed) (Fig. 2). It is recommended to place scanner stations closer to an object in order to get more reliable thermal information. Also, in case of such object, scanner travers with stabilised scan stations of known coordinates would prove to be more effective. They would be more accurate and faster in post-processing and time of field survey would not increase significantly.

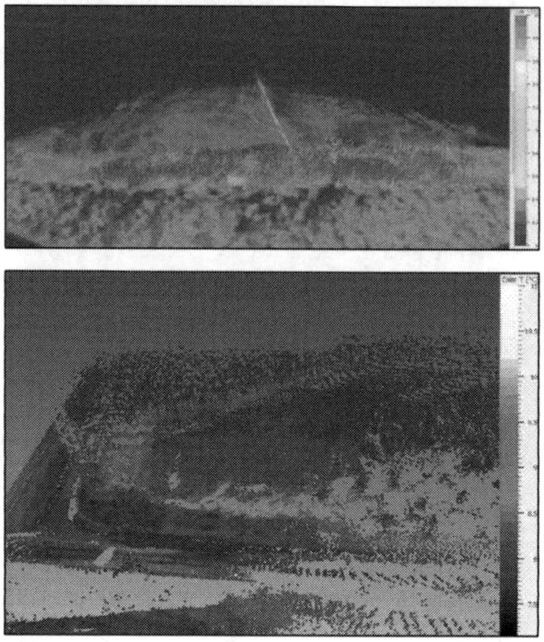

Fig. 2. Top: panorama from Z+F laser scanner with T-Cam Rev 1.0 extension. Bottom: point cloud coloured to some extend with temperature [10].

What is worth to mention that this method is relatively fast during the survey what is extremely important in thermal imaging. Changes in light, temperature of the atmosphere and even amount of water vapour can make images incomparable (if not done in the same conditions an over a short period of time). This is why such survey should take as little time as possible and presented system allows for that.

4 3D inventory of possible voids within spoil tips

None of previously presented survey methods give information on inside structure of the spoil tip. In order to fill this data gap a seismic method was used to monitor the vicinity of the object [6].

Multichannel analysis of surface waves is one of the most popular seismic method for obtaining information on the properties of the subsurface. The method requires a controlled seismic source of energy, such as hammer etc. that induces a surface wave. A properly placed set of geophones, receives generated seismic waves and measures time that waves take to travel from the source (each geophone detects them separately). If the dispersion curve of surface wave is known it is possible to set a process of seismic inversion to build up subsurface model [12], [13], [14].

The acquisition was completed with the use of ABEM seismic system with the use of 24 - 4.5 and 10 Hz geophones placed alternately. Distance between geophones was at 1 m what relates to 24 internal that correspond with 24 metre seismic deployment. At each of them a series of signals in 2 metre intervals was inducted. 7 geophone spreads and 4 response profiles were created. Seismic monitoring was done in three regions, two being flat areas on the top of the object and one on the west slope. On this area thermal monitoring was performed and no fire source or other thermal anomalies were detected. This area was treated as a base of ideally maintained area. It was used as reference area during post processing since there were no potential fire voids detected.

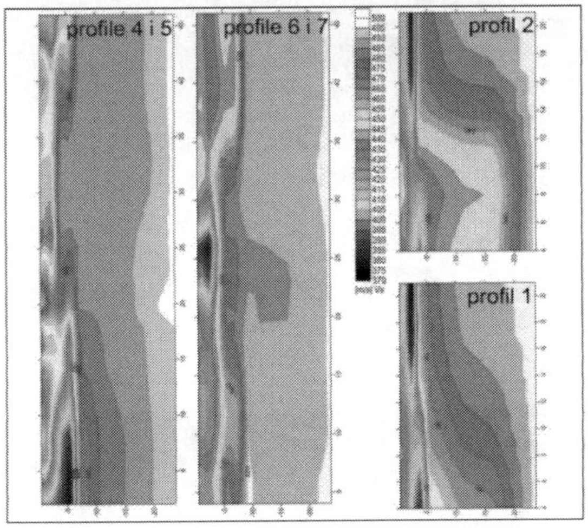

Fig. 3. Multichannel analysis of surface waves – obtained profiles.

Other spreads were placed on the top part of the spoil tip. This area was choose because it was undergoing stabilisation after few years of bulking. The profiles were placed in east-west and north-south direction. The transverse wave speed does indicate that there is a high consolidation of the stored material. During post processing inducted and passive (indicted by passing cargo trains) waves were taken in to account in order to deepen the obtained profile. Results show that material on the top of the spoil tip is less compressed than in lover regions. This is not surprising since bulking process was just finished at the time of the survey. No noticeable voids or significant differences in the kind of stored material were detected. Figure 3 shows the profiles obtained.

Profiles were then connected and a 3D internal structure model interpolated. During acquisition a GPS measurement of geophone placement was performed [11]. It allowed for implementing 3D inside model into Digital Terrain Model thus creating a full 3D model of the structure.

5 Conclusions

The aim of this project was to combine new survey and geophysics technology into one fast method of inside and outside monitoring of coal spoil tips. A monitoring method should be fast, as much work as possible should be done remotely. The final results should be given in a 3D digital forma but also 2D plans and drawings should be given.

The realisation of this project took 5 days of on side work, including gaining access, preliminary walk on the object and gathering existing data from the mine. The field personnel varied but never was more than three people. Only in two instance (2 scan stations of RIEGL scanner and geophysics acquisition) required presence of one or two people on side of the tip. Post processing took about 3 days however it was partly done simultaneously (when geophysics acquisition was being performed scans were registered and processed). The project ended with giving a detail rapport to LW "Bogdanka" environmental division. Personnel was satisfied with quality and quantity of the results.

This project shows that it is possible to combine data obtained by multiple specialists into one 3D model and that with proper planning results could be generate fast. This being important in the view of ensuring public safety and environmental control.

The project was funded by statutory funds of AGH no. 11.11.150.005 and 11.11.190.555.

References

1. J. Amorós López, E. Izquierdo Verdiguier, L. Gómez Chova, J. Muñoz Marí, J.Z. Rodríguez Barreiro, G. Camps Valls, J. Calpe Maravilla, ISPRS Journal of Photogrammetry and Remote Sensing, **66 (1)**, 115 (2011)
2. M. de Berg, M. van Kreveld, M. Overmars, O. Schwarzkopf, *Geometria obliczeniowa. Algorytmy i zastosowania* (Wydawnictwo Naukowo-Techniczne, Warszawa, 2007).
3. M. Cabrelles, S. Galcera, S. Navarro, J.L. Lerma, T. Akasheh, N. Haddad, *Proceedings of the 22nd CIPA Symposium* (Kyoto, Japan, 2009)
4. O. Carpentier, D. Defer, E. Antczak, *QIRT Archives 2004, 088* (2004)
5. J. Drenda, Z. Różański, K. Słota, P. Wrona, Górnictwo i Geoinżynieria **31 (3)**, 149 (2007)
6. A. Kotyrba, M. Grądziel, K. Gogola, Przegląd Górniczy **68 (4)**, 53 (2012)
7. J. Korski, Górnictwo i geologia **5 (2)**, *91* (2010)
8. T. Lipecki, *Kompleksowa ocena stanu geometrycznego obiektów i urządzeń szybowych z zastosowaniem skaningu laserowego* (Wydawnictwo AGH, Kraków, 2013).
9. P. Lewińska, Wiadomości Górnicze **69 (4)**, 79 (2012)
10. P. Lewińska, A. Dyczko, Journal of Ecological Engineering **17 (4)**, 170 (2016)
11. K. Maciuk, Artificial Satellites **51(1)**, 1 (2016)
12. C.B. Park, R.D. Miller, J. Xia, Geophysics **64 (3)**, 800 (1999)

13. C.B. Park, R.D. Miller, J. Xia, *68th Ann. Internat. Mtg. Soc. Expl. Geophys., Expanded abstracts,* 1377 (1998)
14. C.B. Park, R.D. Miller, J. Xia, *Proceedings of the SAGEEP*, SSM4 (2001)
15. R. Welch, M. Madden, T. Jordan, PRS Journal of Photogrammetry & Remote Sensing **57 (1-2)**, 53 (2002)
16. B. Więcek, G. De Mey, *Termowizja w podczerwieni. Podstawy i zastosowania* (Politechnika Łódzka, Łódź 2010)
17. S. Wasilewski, T. Chrobak, Górnictwo i Geologia **35 (1)**, 93 (2011)
18. W. Yun-jia, S. Yao-bin, G. Qiang, S. Yue-yue, W. Xiu-jun, Z. Zhi-jie, *The International Archives of the Photogrammetry, Remote Sensing and Spatial Information Sciences* **XXXVII (B8)**, 203 (2008)

The use of GIS tools for road infrastructure safety management

Marcin Budzyński[1,*], *Wojciech* Kustra[1], *Romanika* Okraszewska[1], *Kazimierz* Jamroz[1], and *Jerzy* Pyrchla[2]

[1] Gdansk University of Technology, Faculty of Civil and Environmental Engineering, Department of Highway Engineering, Narutowicza St. 11/12, 80-233 Gdansk, Poland
[2] Gdansk University of Technology, Faculty of Civil and Environmental Engineering, Department of Geodesy, Narutowicza St. 11/12, 80-233 Gdansk, Poland

Abstract. There are many factors that influence accidents and their severity. They can be grouped within the system of man, vehicle and environment. The article focuses on how GIS tools can be used to manage road infrastructure safety. To ensure a better understanding and identification of road factors, GIS tools help with the acquisition of road parameter data. Their other role is helping with a clear and effective presentation of risk ranking. GIS is key to identifying high-risk sections and supports the effective communication of safety levels. This makes it a vital element of safety management. The article describes the use of GIS for the collection and visualisation of road parameter data which are not available in any of the existing databases, i.e. horizontal curve parameters. As we know from research and statistics, they are important factors that determine the safety of road infrastructure. Finally, new research is proposed as well as the possibilities for applying GIS tools for the purposes of road safety inspection.

1 Introduction

In the last decade nearly 65,000 people were killed and more than 700,000 were injured in Poland's road accidents. Each year more than 70,000 road users fall victim to road traffic. World Bank experts estimate that each year Poland loses 2.7 % of its GDP due to road accidents. Despite a clear reduction in fatalities in recent years, about 3,000 people are killed in accidents on the entire road network. With a demographic rate (number of fatalities per population) of 10, Poland is the EU's worst performing country. There are many factors that influence accidents and their severity. They can be grouped within the system of man, vehicle and environment. The article focuses on how GIS tools can be used to collect, analyse and visualise roadside factors which may lead to accidents. To ensure a better understanding and identification of road factors, GIS tools help with the acquisition of road parameter data. Their other role is helping with a clear and effective presentation and communication of risk and risk ranking.

2 GIS in road transport

In the early 1960s GIS technology was used to create digital cartographic resources offering a spatial presentation of specific features. As the system evolved, GIS was able to present the dynamics of phenomena in terms of time and space [1]. It is based on the concept of overlaying different mapped features on top of each other to determine patterns and causes of spatial phenomena [2]. A geographic information system (GIS) by linking attribute data to spatial data can now be put to use in a number of fields, including transport.

Many countries utilize Geographic Information Systems (GIS) to manage existing transport resources and enhance the effectiveness and efficiency of transport operations. Most researchers found that GIS brings efficiency in data handling and effectiveness in data analysis. The development of GIS applications to road accident analysis has also become a trend [3]. GIS enables a new method in organizing geographical data through its database management system that facilitates the storage, organization and retrieval of digital records [4]. The literature includes a number of examples of how GIS can be used in road safety management [5][6][7][8].

As regards road transport, GIS applications range from simple maps presenting a specific feature (number of cars, accidents, road density, etc.) in terms of space to maps presenting the time and space dynamics of how certain phenomena change. The tool is commonly used for visualising transport data which are collected for official databases such as Central Statistical Office databases (fleet size, fuel prices, etc.) or SEWIK (Accident and Collision Reporting System, number of accidents to identify black spots or black zones) [7]. GIS can be used for accident analysis to examine spatial characteristics of accident locations. Point-in-polygon overlays link the characteristics of the surroundings to the characteristics of the accident. GIS is used as a tool for studying the relation between factors or the effect of factors and events and processes such as the relationship between average speed, speed variation, and accident rates [6] where GIS tools are used to achieve the appropriate data and derive the weight matrix among

* Corresponding author: marcin.budzynski@pg.edu.pl

 https://doi.org/10.1051/e3sconf/20182600009

neighbouring segments which is necessary for the spatial model. More complex analyses, e.g. the effects of characteristic factor features (roadside, type of road surface, road geometric parameters, etc.) on the occurrence of an incident are made possible by point-on-line overlays in which attributes of the road infrastructure can be added. Analyses such as these are an important part of risk management (Fig. 1). The role of risk assessment is to improve road transport safety and define the priorities for public health and public and individual safety [9].

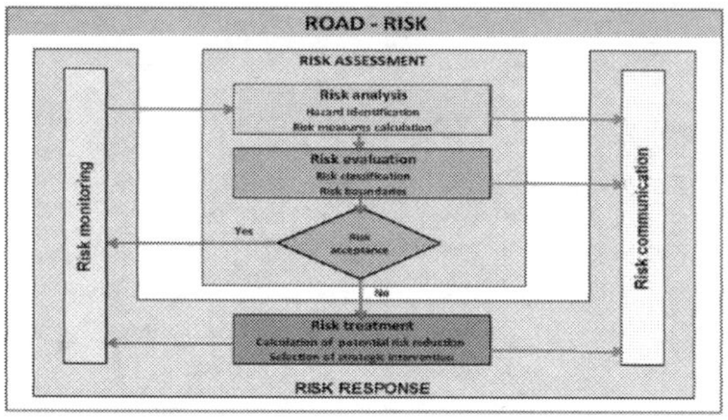

Fig. 1. Diagram of road risk management [11].

In Poland GIS tools are used for safety management at two levels. The first level is designed to identify the exact location of accidents (the Police have only been using GPS for this purpose for three years) and present the effects of road accident database exploration by visualising them on dedicated maps (of countries, regions, counties) of road sections. The result is a clear depiction of different types of risk rankings by area, road network or section.

The second level is designed to collect and visualise road parameter data which are not normally covered in any of the available databases. Tools are needed to ensure that road factor data can be sourced effectively for inclusion in road safety models. These include GIS tools because they can acquire data from the road network automatically, especially as regards geometric parameters.

3 Road infrastructure safety management (RISM)

The objective of road safety infrastructure management (RISM) is to ensure that when road infrastructure is planned, designed, built and used road risks can be systematically identified, assessed for their consequences and eliminated or mitigated. The consequences are measured with the number of accidents, injuries, fatalities and road accident costs. To build RISM tools, we can build on the requirements set out in Directive 2008/96/EC [10]. It covers both planned and existing road networks. Poland introduced these requirements a few years ago.

Because GIS is vital to identifying high-risk sections and effectively communicating safety levels, it plays a key part in the process of safety management.

4 Examples of gis applications

Poland's first application of risk management as part of highway engineering goes back to 2006, when it joined a collaboration with EuroRAP (European Road Assessment Programme). Next, in 2008 as work on the Road Safety Programme for National Roads until 2013, "National Roads GAMBIT" progressed, a suggestion was made to base national roads safety management on risk management methods [12]. Risk maps attracted increasing interest from the media, politicians and road users and were made available on Polish and EuroRAP websites. The maps are still in use today. They inform motorists about the risk on the road network (national and local roads). This helps them choose the lowest risk road and be extra careful when using high-risk sections. In addition, the maps help road authorities with making informed decisions about possible safety treatments. The maps also highlight the differences in safety levels across Poland's national roads and in other countries.

Poland did not do well in such comparisons. Because the results were published at the European forum, road authorities and traffic managers were forced to take more effective action to reduce risk on national roads. Fig. 2 shows a comparison of individual risk maps (measured as concentration of serious accidents) for national roads. The progress of recent years is clear with fewer very high-risk (black sections) and high-risk sections on national roads (red sections).

Fig. 2a. Maps of individual risk on the network of national roads: 2005 – 2007 years [18].

 https://doi.org/10.1051/e3sconf/20182600009

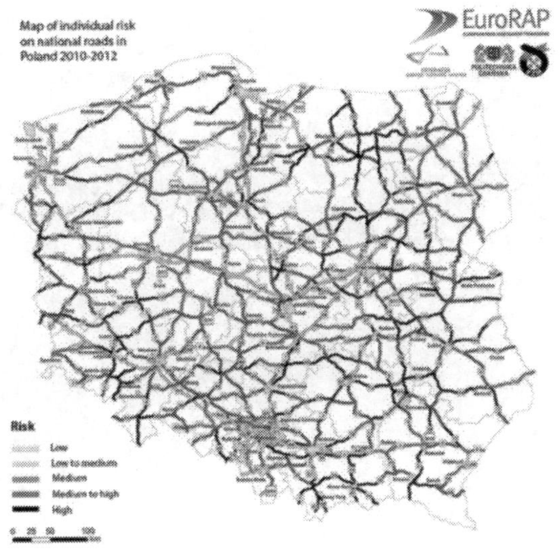

Fig. 2b. Maps of individual risk on the network of national roads: 2010 – 2012 years [18].

Because fatalities on regional roads are much lower compared to national roads, the fatality criterion of road safety is simply not enough. Societal and individual risk is assessed by studying road deaths and serious injuries. This approach is in line with the EU's strategy [13] and Poland's 2013 National Road Safety Programme for the Years 2013-2020 [14]. Fig. 3 shows individual risk in terms of accident cost concentration (accident costs versus kilometres travelled) on regional roads and Fig. 4 shows societal risk (fatalities and serious injuries) by poviats. The data cover the period between 2012 and 2014 [15].

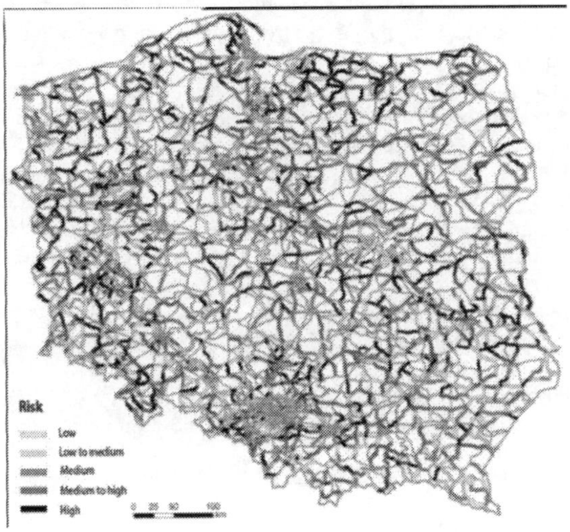

Fig. 3. Map of individual risk on regional roads [15].

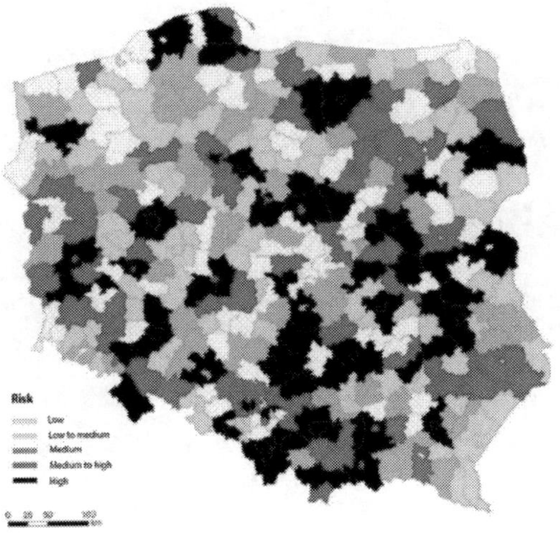

Fig. 4. Map of societal risk by poviats [15].

5 Road data for modelling hazards

One of the key requirements for an effective and efficient management of road infrastructure, both existing and planned, is availability of data on the level of risk and road safety on road networks, sections or junctions. The relevant decisions must be supported with a good understanding of the current road safety and the hazard that may impact safety over the coming years. [16], [17].

With access to data on accidents and traffic, we can calculate the risk measures which will help us to identify safety levels on selected elements of the road. Sections can then be identified where accident risk, fatalities and injuries are the highest [18]–[20].

Key to safety management is verifying the available data. As an example, by carefully verifying SEWIK data for the region of Warmińsko-Mazurskie in 2013 [21], it was possible to improve the accuracy of road category data by 30 % (Fig. 5). This was achieved by using road records, GPS coordinates and other information.

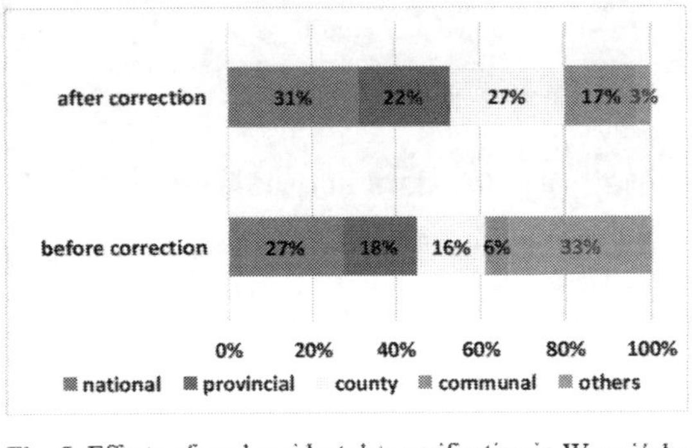

Fig. 5. Effects of road accident data verification in Warmińsko – Mazurskie [21].

Effective road safety management is also ensured by using data on elements of the roads. The set of independent variables that have an effect on the number and variability of section-related road safety measures is quite broad. Road safety models are built using variables of road horizontal alignment in connection with horizontal curves.

6 Parameters of horizontal curves

Up until now Poland's models of road safety measures have not included the parameters of horizontal curves because the relevant data are difficult to obtain. While a large part of the models contain variables that include road geometry parameters, it is not clear how they impact road safety [22]. This was addressed in previous research [22], but the samples were not big enough. Road curvature is included in models that represent long sections (10-50 km) [23]. Parameters of horizontal curves are particularly important in Poland due to a high number of fatalities involved in run-off-road accidents. It is estimated that app. 40 % of deaths involved in run-off-road crashes occur within the area of horizontal curves. It is clear from the statistics how serious the problem is – a combination of horizontal curves with dangerous roadsides. With app. 20 % of all fatalities being the result of run-off-road accidents, mitigating this hazard becomes a priority. Horizontal curves and their parameters have a strong impact on road safety [24][25][26][27][19][28][29]. Horizontal curves are strictly related to speed and visibility [30]. Speed profile, profile and curvature of vehicle trajectory on horizontal curves are affected by road qualitative features such as radius, angle of turn, width of road and road surface condition as a result of the weather (dry and wet road surface) [31]. Limited sight distance involves hazards such as junctions within horizontal curves, elements of road interchanges (ramps), pedestrian crossings, etc. There may be additional hazards as a result of the demand for overtaking when speed dispersion is high. If sight distance is limited, this may lead to head-on collisions.

All these hazards and statistics show that on average in a year app. 10 % of all accidents and 14 % of all fatalities on Polish roads occur on horizontal curves (app. 400 fatalities). Because it is such a serious problem, road safety models must be enhanced to include horizontal curves.

7 Methods for data acquisition

Research conducted by the authors shows that there are other factors that affect safety on specific road sections which have not been studied in-depth so far. These factors include the number of horizontal curvatures as a sequence of connected curves, horizontal geometric parameters of curves, direction of movement on the curve and the total length of the curvature of curves.

To determine how these problems affect road safety, road sections have been selected for in-the-field research (national road no. 20, pomorskie region, lenght 20 km).

One strand of the authors' work is building and testing a system for an effective horizontal curve data acquisition. The other equally important strand is analysing the data for use in the process of road safety model design. This should help select the right treatments.

Road sections were selected for data registration using the designed system. The results were then verified on those sections [32]. Fig. 6 shows a test section.

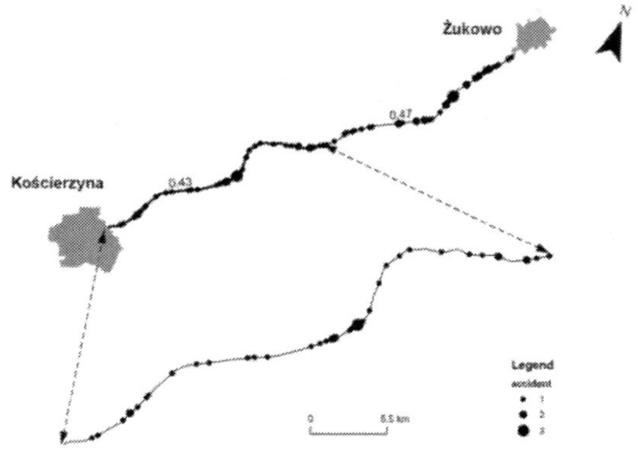

Fig. 6. A road section which was part of the inventoried measurements [32].

The experimental apparatus consists of two types of sensors. The first group is made up of three integrated digital cameras. The second includes a mobile RTK receiver (Leica Viva CS15) with a built-in GSM to ensure access to corrections made available by RTK reference stations. During the experiments adjustments were sourced from the ASG – EUPOS network. The positions registered by the receiver were transmitted in real time to a portable computer with ArcGIS software. Once collected, the data were checked for technical correctness. The checks looked at horizontal coordinates and registration errors. The idea was to determine whether the registered points put on the road's current orthophotomap are within its boundaries or not.

The data were then used to determine curve sections and analyse the parameters of the curves. The basic parameter of each curve was its length and whether it linked into a subsequent curve. The number of accidents on a particular road section provided a good background for its characteristics.

Road curvature analysis helped to develop an algorithm for determining road sections considered curves and their parameters, i.e. curve radius and curve length. The point of departure was the assumption that curvature length is an important road safety factor.

8 Conclusion

GIS tools support the acquisition of road parameter data. So far Poland's road safety models have not covered selected elements of road network such as horizontal curves. By using GIS we can present risk ranking in a clear and effective manner and better communicate the risks. This makes the tools a vital element of road infrastructure safety management. The next stage of the research and GIS tools application will be to use them in road safety inspection procedures, which is one of the road infrastructure management procedures that Poland implemented three years ago. This needs further development and improved effectiveness and efficiency.

Finally, the research should end in building a comprehensive model of the effects of road parameters on road safety measures using GIS techniques.

References

1. M. Kuźma A. Stachowicz, Rocz. Geomatyki **14 (3)**, 387 (2016)
2. M. Peggion, A. Bernardin, M. Masera, *Geographic Information and Risk Assesment* (Luxembourg: Office for Official Publications of the European Communities, 2008)
3. K. Lupton, D. Bolsdon, Comput. Environ. Urban Syst. **23 (5)**, 383 (1999)
4. I. Heywood, S. Cornelius, S. Carver, *An Introduction To Geographical Information Systems (3rd Edition)* (Pearson, 2006)
5. K. Austin, M. Tight, H. Kirby, Transp. Plan. Technol. **20 (3)**, 249 (1997)
6. M. Quddus, J. Transp. Saf. Secur. **5(1)**, 27 (2013)
7. T. Steenberghen, T. Dufays, I. Thomas, B. Flahaut, Int. J. Geogr. Inf. Sci.**18 (2)**, 169 (2004)
8. P. C. Lai and W. Y. Chan, Ann. GIS **10 (1)**, 58 (2004)
9. A. S. Hakkert, L. Braimaister, and I. van Schagen, *Proc. Eur. Transp. Conf. 2002* (Association for European Transport AET, London, 2002)
10. The European Parliament and the Council of Europe, *DIRECTIVE 2008/96/EC on road infrastructure safety management* (2008)
11. K. Jamroz, M. Budzyński, W. Kustra, L. Michalski, S. Gaca, Transp. Res. Procedia **3**, 730 (2014)
12. K. Jamroz, S. Gaca, J. Gasparski, *Road Traffic Safety Program for National Roads until 2013* (Association of Communications Engineers and Technicians of Poland, 2009)
13. European Commission, *Road safety program 2011-2020* (2010)
14. KRBRD, *National Road Safety Program 2013-2020* (2013)
15. K. Jamroz, W. Kustra, J. Wachnicka, and M. Berkowski, *Risk classification methodology for selected types of road accidents on voivodeship roads as well as for voivodeships and poviats, including the classification and presentation of results on maps* (Ministry of Infrastructure and Development Secretariat of the National Road Safety Council, 2015) (in Polish)
16. K. Jamroz, *Risk management method in road engineering* (Wydawnictwo Politechniki Gdanskiej, 2011)
17. K. Jamroz, L. Smolarek, Int. J. Occup. Saf. Ergon. **19 (2)**, 297 (2013)
18. M. Budzyński, K. Jamroz, W. Kustra, S. Gaca, L. Michalski, *Classification of dangerous road sections on national roads GDDKiA* (Gdansk University of Technology, Cracow University of Technology, 2013)
19. D. W. Harwood, F. M. Council, E. Hauer, W. E. Hughes, A. Vogt, *Prediction of the Expected Safety Performance of Rural Two-Lane Highways* (2000)
20. EuroRAP, *Road protection score (RPS) method and pilot results*, (2002)
21. J. Żukowska, T. Radzikowski, Transp. Miej. i Reg. **(4)**, 31 (2004)
22. W. Kustra, *56 The Scientific Conference of the Committee of Civil Engineering and Water of the Polish Academy of Sciences and the Science Committee of the Polish Academy of Sciences*, 329 (2010) (in Polish)
23. M. Budzyński, *Characteristics of rural roads and the threat to health and life of their users*, PhD thesis (2005)
24. F. Bella, IATSS Res. **39 (1)**, 51 (2015)
25. S. Mondal, Y. Lucet, W. Hare, Comput. Oper. Res. **64**, 130 (2015)
26. P. Greibe, Accid. Anal. Prev. **35 (2)**, 273 (2003)
27. G. Khan, A. Bill, M. Chitturi, D. Noyce, Transp. Res. Rec. J. Transp. Res. Board **2279**, 124 (2012)
28. F. Bella, Transp. Res. Procedia **3**, 100 (2014)
29. C. Wang, M. A. Quddus, S. G. Ison, Saf. Sci. **57**, 264 (2013)
30. J. Kempa, *Characteristics of free movement on curvilinear stretches of roads in the plan* (Wydawnictwo Uczelniane Uniwersytetu Technologiczno-Przyrodniczego, Bydgoszcz, 2008) (in Polish)
31. P. Buddhavarapu, A. Banerjee, J. A. Prozzi, Accid. Anal. Prev. **52**, 9 (2013)
32. M. Budzynski, K. Jamroz, J. Pyrchla, W. Kustra, K. Pyrchla, *2016 Baltic Geodetic Congress (BGC Geomatics)*, 154 (2016)

 https://doi.org/10.1051/e3sconf/20182600010

A concept for the modernization of underground mining master maps based on the enrichment of data definitions and spatial database technology

Artur Krawczyk[1,*]

[1]AGH University of Science and Technology, Faculty of Mining Surveying & Environmental Engineering; Department of Mine Areas Protection, Geoinformatics and Mine Surveying, al. Mickiewicza 30, 30-059 Cracow, Poland

Abstract. In this article, topics regarding the technical and legal aspects of creating digital underground mining maps are described. Currently used technologies and solutions for creating, storing and making digital maps accessible are described in the context of the Polish mining industry. Also, some problems with the use of these technologies are identified and described. One of the identified problems is the need to expand the range of mining map data provided by survey departments to other mining departments, such as ventilation maintenance or geological maintenance. Three solutions are proposed and analyzed, and one is chosen for further analysis. The analysis concerns data storage and making survey data accessible not only from paper documentation, but also directly from computer systems. Based on enrichment data, new processing procedures are proposed for a new way of presenting information that allows the preparation of new cartographic representations (symbols) of data with regard to users' needs.

1 Introduction

The first use of databases in the underground mining industry in Poland dates back to the 1980s. At that time, they were mostly used in financial management and bookkeeping tasks. It was not until the 90s that CAD (Computer Aided Design) was implemented in the industry for the purpose of creating digital mining maps. This involved implementing software not only for mine scheduling and management of geological resources, but also in various other mining departments. Since this process was not sufficiently controlled, some issues with sharing spatial data stored in binary CAD files (AutoCAD, MicroStation) arose.

These problems can be divided into two categories: 1) parallel sharing and viewing of mining map files and 2) limitations in controlling and managing all aspects of geological and survey records. The problem with the "parallel sharing" of data by many users comes from the design of CAD technology. Vector data stored in binary files is not designed for simultaneous use by many users. Map data files have to be updated to the current state of excavation at the end of each annual quarter. The number of employees needed for this update could exceed the number of installed copies of CAD software as mining companies usually buy as few licenses as possible due to the expense. This causes problems with accessing the data. The data management limitations are caused by the complex way in which survey and geological data are organized. This includes handmade sketches on paper and the results of calculations done in various survey programs which are printed and saved in binders stamped and signed by an authorized employee. During the map creation process in CAD, the vector graphic system loses attribute data, which remain only in paper form. The scope of data on mining maps was defined in the 70s and is no longer sufficient. A lot of data has to be interpreted from original two-dimensional maps in order to extrapolate spatial information about the third dimension [1].

2 Some details on underground mining maps

The definition of an underground mining map states that it should present the current mining situation and geological structure. It also should include the condition of the terrain surface (with buildings) within the borders of the mining area. In order to fulfill these conditions, each mine needs to have surveying and geological departments which prepare and review master maps and special (detailed) maps. In Poland master maps are usually created at a scale of 1:2000, while review maps use a scale of 1:5000 or 1:10 000. Master and review maps detail the actual conditions in the mine. In turn, special maps show conditions in a different color, or simplifies them in order to enhance the primary purpose of the map.

* Corresponding author: artkraw@agh.edu.pl

 https://doi.org/10.1051/e3sconf/20182600010

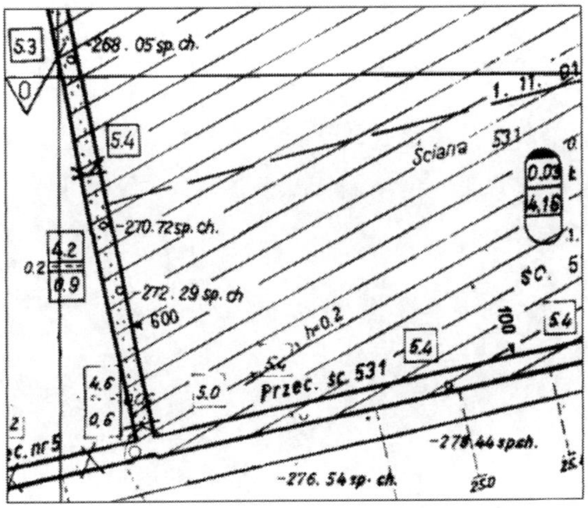

Fig. 1. A section of an underground mining map (1:5000).

The creation of underground mining maps consists of a few stages. Firstly, a survey is done and the results are written accordingly in survey and geological books. All calculations are done in appropriate software and stored. The third and most important part is drawing the map. This process can be done in the traditional manner by hand-made drawings on paper maps glued to an aluminum plate, or it can be done in CAD software enhanced by dedicated plugins.

It is important to emphasize that the visual part of a map is a highly sophisticated and complicated product; therefore, special software plugins for drawing must be used.

The regulations regarding the described maps are given in Geology and Mining Law [2] Article 116, which states that it is necessary to have survey and geological documentation that should be updated in relation to the excavation process. This documentation includes:

1. survey reports,
2. calculation reports (usually provided by survey software),
3. cartographic documentation of the current geological and mining situation, including the surface condition (and buildings) within the borders of the mining area.

Article 116 specifies that only qualified surveyors and geologists are allowed to provide survey and geological information and that they are obligated to update documentation with appropriate survey data and calculations. The "Geological and Mining law" [2] regulation in Article 116 refers to the Regulation of the Ministry of Environment dated 22 December 2011, regarding survey and geological documentation. Paragraph 7 point 2 describes the conditions that cartographic digital documentation must fulfill. Digital cartographic documentation can be done if:

1. it is properly secured,
2. all updates are stored and recorded so that older versions of files can be restored,
3. the mining company can plot the digital document at any time. If this is not done, the mining company must store a plot of an updated version of the documentation.

On this basis, digital maps can be implemented only if:

• Original maps are stored in servers managed and located in survey departments.
• Digital copies of the maps are stored in two places: one outside the mine (e.g. in the central office) and the other on the server in the mine survey department.
• A paper map repository stores paper plots of digitally updated maps instead of manually updated paper maps.
• The plots are done every 3 months; only updated sections are plotted.
• Sections not updated need to be plotted every year.

This last condition relates to the physical properties of paper, which can deteriorate with time. In uncontrolled conditions, maps lose the cartographic properties within 6 months. In air-conditioned conditions, printed maps can remain usable for up to a year. Based on this regulation, it is possible to stop creating and manually updating paper on aluminum plate maps by converting these maps to digital versions.

All these regulations mean that the primary task of the survey department is to obtain a product in accordance with the legal regulations applying to the storage.

The main problem with new CAD solutions for creating and plotting maps is that attribute (textual) information is still not attached to the graphic elements.

The basic information provided by the survey department is the geometry of sidewalls measured by survey mining staff. The xyz geodetic coordinates of these points are then calculated and the points are mapped and drawn as continuous lines, but information about the third coordinate (z) is not stored [3]. The original maps are made at a scale of 1:2000 or 1:1000 (depending on the type of mine). When maps are scaled to 1:5000 or 1:10000, they are subjected to cartographic editing that results in descriptions being moved or elevation points deleted due to the density of the information. On the master map, two or three points per 100 m of excavation are drawn as points with ordinate information (the z coordinate value is added as text).

After master map generalization to convert to a scale of 1:10000 or 1:5000, the number of points with a known z value decreases. On the other hand, elevation information is very important for the other departments because it allows the specification of parameters such as the direction of the excavation slope or the slope value. This parameter is also important for underground transport and ventilation purposes. In turn, the height of the excavation is important for geologists and miners.

The ideal situation would be if a survey department could provide all of this kind of data to other departments for them to use freely. This however, cannot be achieved without changes in current data processing.

3 Proposed solutions for reducing limitations on managing survey data for other mine maintenance departments

Paper reports are produced during the creation of surveys and geological documentation. They include handmade survey sketches, miner's sketches and other documentation. The aim of modernization is to convert this paper-based data to a digital format. At the time of this article, no Polish mines have a digital system for storing this kind of data.

The first modernization concept is the "flat solution", which involves scanning sketches and storing them in pdf format with geo-reference data added to the coordinates system used in each mine. The sketches should then be equipped with the results of calculations obtained from specialized survey programs (also as pdf files). The created set of pdf files with measurement information can be referenced by linking to the vector geometry of the sidewalls line drawn on the map. This solution can be used in CAD and GIS environments and improves access to the raw data; however, data in a pdf file is not very convenient for other mining departments.

The second most advanced modernization concept is the "full numeric approach". This idea is based on the modernization of survey data acquisition methods (first stage of map creation) during measurement, such as the use of electronic devices like laptops or tablets instead of paper. Numeric and geometry data from measurements can be stored and transferred from devices directly into numeric spatial database fields such as SDO_Geometry or Well-Known Binary, both of which are commonly used in open source software.

However, using electronic devices underground can cause problems with the flammable methane that is often present in mines. All electronic devices have to be constructed and tested for safety under these conditions. So far, only expensive geodetic instruments have been constructed and certified for use as such.

The least advanced solution is based on the idea of "enriching the map definition" beyond the legal regulations. The solution begins from extending the number of data sets measured during underground surveys. New data sets should be separated from other legal data sets, but they can still be used by other departments. Also, cartographic presentation can be changed depending on the mine maintenance department's needs. The same information should be prepared in a different form for appropriate maintenance departments. Cartographic representation of new data sets leads to the next problem of representation of these data on maps in multi-resolution environments. This involves problems with creating generalization rules when concerting to other scales. These considerations are similar to the proposed solution to problems with generalization in topography maps [4].

4 Modernization technology for survey coordinate data based on enriched map definitions

During map creation by mining surveyors, only a few elevation points inside the excavation are placed on the map, Fig. 2a. The scale of this representation is based on the scale of master maps used in a particular mine. The number of measurements taken during a survey is much higher than is placed on the map. This is done according to the cartographic rules for the creation of mining maps.

The first concept for the "enriched map definition" is based on increasing the amount of data placed on the map. Fig. 2b shows an increased number of elevation points.

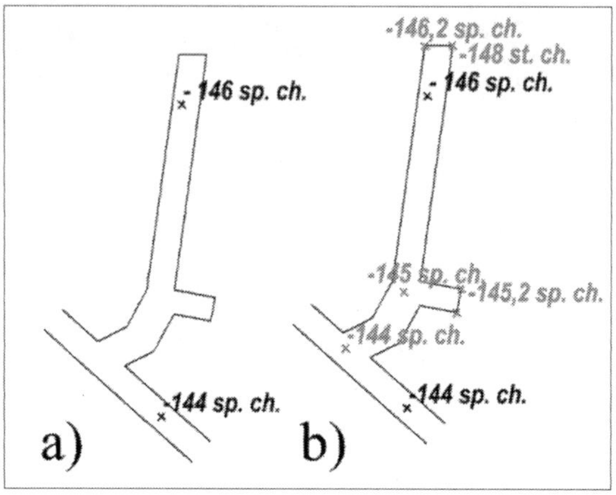

Fig. 2. Geometric representation for a survey: a) typical representation of sidewalls and elevation pegs on the master map; b) proposed additional set of data as "enriched map definition".

These added points are selected according to some rules: 1) for the modeling of spatial mining data, it is very useful to have elevation points on excavation crossings, 2) for geology and mining planning, it is useful to have data about the height of the mining face, 3) for transportation purposes, it is useful to have more information about the direction and value of the excavation slope.

New elevation data presented in Fig. 2b. should be placed in separate data sets with new data attributes, such as the type of data source enrichment. There should not be a cartographic version of this additional data because there is no intention of using it when plotting maps. Elevation data is used to create new information described in the second part of the concept.

The second part of the modernization concept is the creation of new information and cartographic representations for use by other departments:

1) For transportation and ventilation, the key information on the map is the slope of the excavation. These data are needed for planning and servicing transportation systems. The ventilation dept. provides information on the placement of ventilators, ventilation dams, and the direction of the air flow in a mine. The new slope representation is based on the enriched survey data set. The example slope value is placed below the excavation. Units are given in promille ‰ (one-tenth of a percent), or as a percentage. Calculating a slope value requires at least two elevation values from pegs. The first peg must be located at the beginning of the excavation; the second must be located at the end of the excavation. Both pegs must be located on the floor of the excavation (in Polish, pegs labeled "sp. ch." are on the excavation floor). The cartography symbol for the slope is presented in Fig. 3 as an arrow with the value of the slope.

 https://doi.org/10.1051/e3sconf/20182600010

2) For the geology department and the mine casing design department, the key information about the excavation is the floor slope and the height of excavation. For geological reasons, the height is important in order to control the data included on the map. The map contains detailed information about the thickness of geological layers identified in this excavation in meters, but there is no information about the height of the excavation. The total thickness of the geological layers in the excavation should be equal to its height. The height of the excavation is also important for the casing design for new mining works. Before a project starts, the planning stage needs data for evaluation and information about the conditions for the planned mining work. A typical way of obtaining this information is to calculate the height from the floor pegs and the roof pegs (labeled "st. ch."). The problem with this calculation is that the pegs on the roof are usually placed in different locations than the floor pegs, sometimes with remarkably long distances between them. In ideal conditions for calculating, the planar positions (xy) of those pegs should be the same. In the new concept, data filtering provides proper data to obtain the height of the excavation. In Fig. 3, new cartographic symbols are presented for symbolizing the height of the excavation; these are marked as "w. ch." and show the height in meters.

3) For spatial underground modeling, the key information is the axis of the existing excavation, which is a source for planning new mining works.

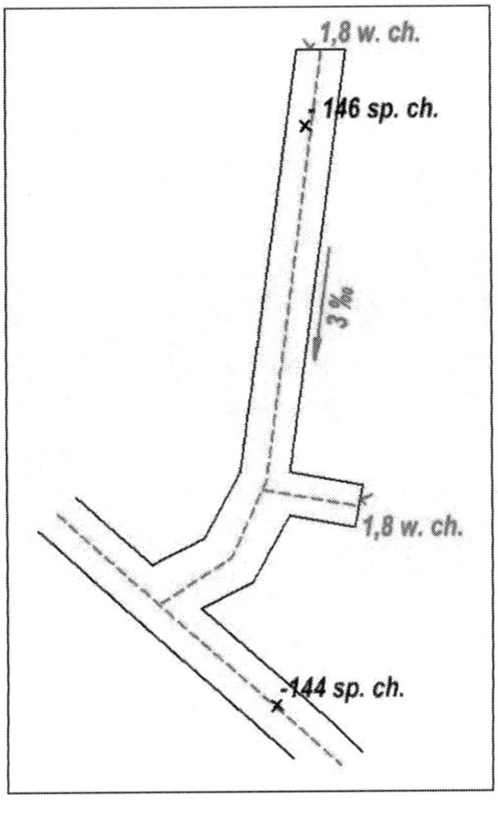

Fig. 3. Enriched cartographic symbols for mining master map.

The first idea of the "enriched map definition" concept indicates that the system should also gather data from the survey for estimation of the axes.

The geometry of the excavation axis shown on the map is a complicated problem in mining information systems. Poniewiera [5] described this problem and implemented a software tool for supporting manual drawing of excavation axes. In the GeoLISP solution, the surveyor manually draws a line between the sidewalls of the excavation. This method works well at a small scale (one person working on around a dozen mining map sheets), but at a large scale (many people, a lot of mining map sheets) the quality of work deteriorates. The primary issue with manual digitizing is mistakes made during hand drawing of the axis when it intersects the line of sidewalls. To avoid this situation, the enriched data model uses more pegs situated in the middle of the excavation floor to prepare a ready-to-use data set for connecting points manually or semi-automatically. The excavation axes play several important roles in spatial data modeling for mining works [6], numeric data modeling of seams [7], and also for visualization and animation [8].

5 Concept of mining base map stored in a spatial database

So far in mining in Poland, spatial database technology has not been used to store and manage legal mining maps. However, [9] in 2013 LW Bogdanka S.A. (a mine in Poland) started planning and implementing a new system called Mining Infrastructure Object Maps (MIOM) that is based on spatial database technology. This system was created for control over assets in underground excavations. It collects information about mining machinery and equipment such as catalog identification number, technical description and location. Spatial data is stored in an Oracle Database, where it is edited by Bentley Map GIS class software. This system was created in long-term cooperation with the LW Bogdanka S.A. Coal Mine and IGSMiE PAN and used for the inventory process of underground infrastructure. In 2010 in the Bogdanka mine, PRGW Company implemented a CAD-based system for mining master map production called SoftMine MapaKWK. This shows that two spatial data systems already exist. Where a master mining map is used as a reference map for a Mining Infrastructure Object Map [10], the cooperation between these systems is possible only because both systems use the same graphic platform: Bentley MicroStation.

This example shows that a spatial Oracle database is well-suited to storing any kind of spatial data [11], including data defined for master mining maps. This system can also store scanned miner's sketches and data from survey instruments. Based on this, the database can also store all input data and result data from survey calculations, therefore this data can be secured and used by miners from other departments. Relational databases provide tools for data sharing and editing in a multi-user environment. Edited data can be viewed but not edited by unauthorized users. Geometry stored in spatial tables can be edited by GIS class software, thus allowing the creation

https://doi.org/10.1051/e3sconf/20182600010

of a master mining map without the use of CAD software. An additional advantage of spatial data bases is high level data interoperation based on markup languages [12]. The main problem with this solution is the separation of geometry data styling, which is not a common technique in the mining industry.

6 Conclusion

The most important objective is to connect two trends into one solution for all mine maintenance departments. The first problem is that the old definition of master mining maps is obsolete and needs to be improved. The second problem is the modernization of technology to allow the transfer of data from CAD systems (editing and storing) to spatial databases. These two trends are interdependent. New technology offers more ways of gathering and analyzing spatial data, and this is exactly what users need.

Of course, the concept proposed in this article needs further analysis in order to determine the details of implementation. It is also important to do further tests on a bigger data set. However, even this generalized concept provides an alternative to the currently used methods of editing and storing surveyed data. Databases represent an excellent tool for complying with the new laws that are pending in the mining industry.

The article was done in the framework of the "Badania Statutowe" grant of the Department of the Protection of Mining Areas, Geoinformation and Mine Surveying, AGH-UST no. 11.11.150.195

References

1. L. Wu, Z. Yin, Y. Zhong, Journal of China Coal Society **28 (1)**, 1 (2003)
2. Geology and Mining Act (Journal of Law 2011/163/981, 2011)
3. J. Wang, B. Yang, M. Wu, Coal Science and Technology **32(5)**, (2004)
4. K. Kozioł, A. Krawczyk, M. Lupa, *Beyond Databases, Architectures, and Structures*, 435 (Springer International Publishing Switzerland, 2014)
5. W. Muras, M. Poniewiera, Zeszyty Naukowe Politechniki Śląskiej: Górnictwo **1752 (278)**, 313 (2007)
6. J. Maciaszek, R. Gawałkiewicz, I. Gawałkiewicz, Zeszyty Naukowe AGH: Geologia **36 (3)**, 331 (2010)
7. M. Poniewiera, Wiadomości Górnicze **61 (7–8)**, 458 (2010)
8. A. Krawczyk, *VII Konf. Nauk.-Tech. Ochrona Środowiska na Terenach Górniczych*, 127 (Katowice, Poland, June 2008)
9. S. Karlikowski, T. Parzniewski, D. Galica, *Materiały Konferencyjne SEP* (Cracow, Poland, 2014)
10. T. Parzniewski, D. Galica, A. Dyczko, A. Krawczyk, Wiadomości Górnicze **67 (2)**, 115 (2016)
11. A. Godfrind, R. Kothuri, E. Beinat, *Pro Oracle Spatial for Oracle Database 11g* (USA, 2007)
12. A. Krawczyk, Gospodarka Surowcami Mineralnymi **23 (4)**, 163 (2007)

As- built inventory of the office building with the use of terrestrial laser scanning

Marek Przyborski[1], *Paweł* Tysiąc[1,*]

[1]Gdansk University of Technology, Faculty of Civil and Environmental Engineering, Department of Geodesy, Narutowicza St. 11/12, 80-233 Gdansk, Poland

Abstract. Terrestrial Laser Scanning (TLS) is an efficient tool for building inventories. Based on the red-laser beam technology it is possible to provide the high accuracy data with complete spatial information about a scanned object. In this article, authors present the solution of use a TLS in as-built inventory of the office building. Based on the provided data, it is possible to evaluate the correctness of built details of a building and provide information for further construction works, for example an area needed for Styrofoam installation. The biggest problem in this research is that an error which equals over 1cm could generate costs, which could be a problem to cover by a constructor. Based on a complicated place of the construction works (centre of a city) it was a challenge to maintain the accuracy.

1 Introduction

In according to the polish and international literature it is needing to perform a proper inventory of buildings [1][2]. Geodetic measurements are present on every step of buildings construction stages from map to design proposes to as-built inventory. Crucial is to maintain a proper accuracy of the data in every of stage.

Nowadays, one of the option for fast and accurate data collecting is the use of laser scanning technology. Based on the red- beam laser technology could provide a spatial information about the object, which is represent by millions of points. The most challenging is to operate on such data which require lots of computing power by hardware. It could be a disadvantage, but it has to be mentioned that point cloud is not a final product. From point cloud, for example CAD drawings could be generate and to operate on such data is not a problem.

One of the advantage of use such a technology is that besides of the information about coordinates of points, the amplitude, reflectance values are registered. Reflectance of the surface could be defined as its effectiveness in reflecting a radiant energy from that surface. This application is reliable in estimate a condition of the materials [3][4] and ensure the view of the point cloud in so called 'false colours'. That gives the user opportunity to have a valuable information about object's surfaces.

Moreover, in describing laser scanning technology, we could distinguish three systems: terrestrial, mobile and airborne. Main differences between them are accuracy of the provided data and land they could covered in one measurement mission. For single construction, the best is to use the most accurate-terrestrial laser scanning [5].

2 Mission planning

The object of interest is placed in Gdansk, Poland and is represented by a construction of a new office building. The approximate location with geographical coordinates is presented in Fig. 1 as the print screen from Google Earth software inside the red polygon.

54°21'20'' N 18°35'42'' E

Fig. 1 The area of interest [source: Google Earth]

Assumptions of the whole measurement was to obtain the 1 cm accuracy for checking the correctness of zero state of a building for example, the size of carved out windows and doors holes. What is more, walls deviations were calculated and every of dimensions between windows and doors holes were estimated because of a Styrofoam, which was going to install. Based on the accuracy assumption, the constructor could reduce the cost of the materials. It has to be mentioned

* Corresponding author: pawtysia@pg.edu.pl

 https://doi.org/10.1051/e3sconf/20182600011

that a lot of Styrofoam during the installation is going to the trash. With the use of laser scanning technology, it could reduce not only costs but also the time need for Styrofoam installation.

The laser scanner which has been used in this experiment was the Riegl VZ- 400, provided by Apeks company, placed in Gdansk, Poland. According to the producer brochure about the equipment, Riegl VZ-400 is able to collect the data with 3 mm precision and 5 mm accuracy [5]. The one more problem to investigate in post-processing the data is the level of noise, which has to be filtered before the data processing. Moreover, the Riegl ensure the user that the instrument could collect 122 thousand points per second in High Speed mode. Based on that instrument model and measurement specification, a one scan pattern for each wall is required and analyse separately.

Unfortunately, the situation was different from presumed. The whole object was enclosed by metal plates, preventing from enter into the construction site. Also, a close neighbourhood of road and other buildings was a problem for data acquisition. It was not possible to cover assumptions. It was a need to create a lot more scan stations. Authors placed 15 scan positions. The alignment of that positions and estimate the perfect, according to the project place of walls for deviations calculations was crucial. The Fig. 2 is presenting the scanned object.

Fig. 2. The view of a scanned object.

The Fig. 3 is presenting the places where the instrument measured the object.

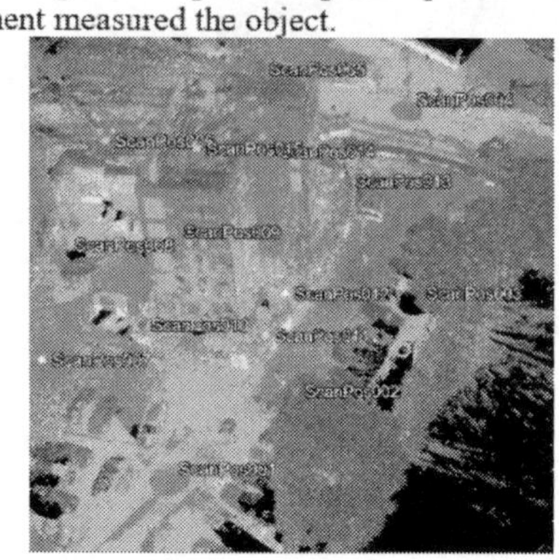

Fig. 3. Positions of the instrument.

3 Alignment and plane estimation

To align scan positions between each other, the Iterative Closest Points method was used. To maintain the accuracy of the data, first it had to indicate the scan positions belonging to one scanned wall. Alignment 15 scan positions could generate an error, much exceeding of the assumed. After the alignment, the value of noise has to be estimated. Whole gained data accuracy has not exceeded 1 cm. The process of alignment evaluate is described below.

First, scan positions number 1, 2 and 7 was assigned to the North wall. Scan positions number 2, 3, 11, 12 and 13 was assigned to the West wall. Scan positions number 4, 5, 14, 15 was assigned to the South wall and 5, 8, 9, 10 to the East wall. The alignment results were presenting below in Fig. 4 for the North wall, Fig. 5 for the West wall, Fig. 6 for the South wall and Fig. 7 for the East wall. Every of standard deviations results for each wall are presenting in meters.

Fig. 4. The alignment results for the North wall.

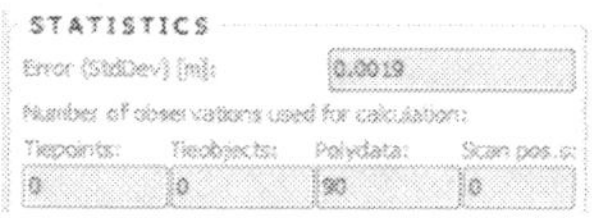

Fig. 5. The alignment for the West wall.

Fig. 6. The alignment for the South wall.

https://doi.org/10.1051/e3sconf/20182600011

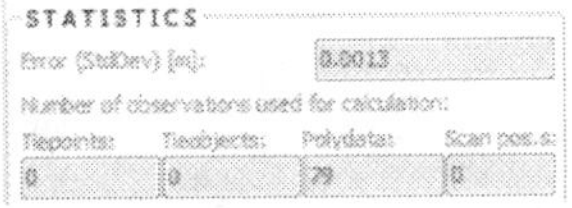

Fig. 7. The alignment for the East wall.

The plane was estimated with the use of the least square- method by indicate 8 points, belonging to the walls at the bottom of walls. The representation of chosen points is presenting in Fig. 8. The same workflow was performed for each wall. The reason for placing points was that from the level 0 to the height of 1 meter it could be observed the hypothetical, non – angular deviated walls of the building. From that wall (It could be named- reference), the angular, real deviation was estimated.

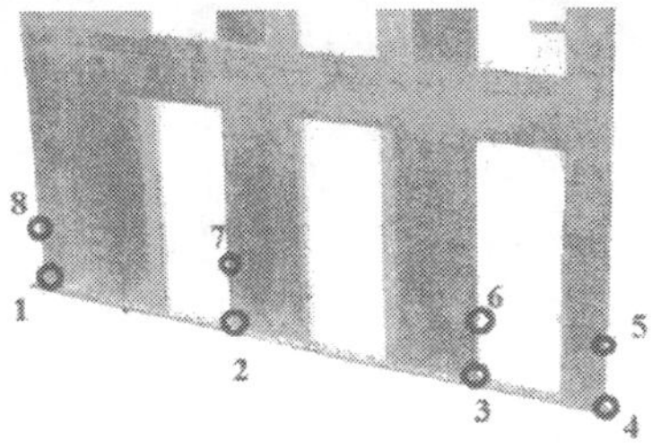

Fig. 8. Chosen points for plane matching which indicate the wall according to its perfect position.

4 Experiment results

First, dimensions of windows, doors gaps were measured and comparison between measured ones and from an architectural plan sizes was performed. The results on the example of North wall (down part of the wall, the same as its seen in Fig. 8) was presented in Fig. 9. It could be noticed that the differences between sizes are significant and up to 3.5 cm with the accuracy of 1 cm.

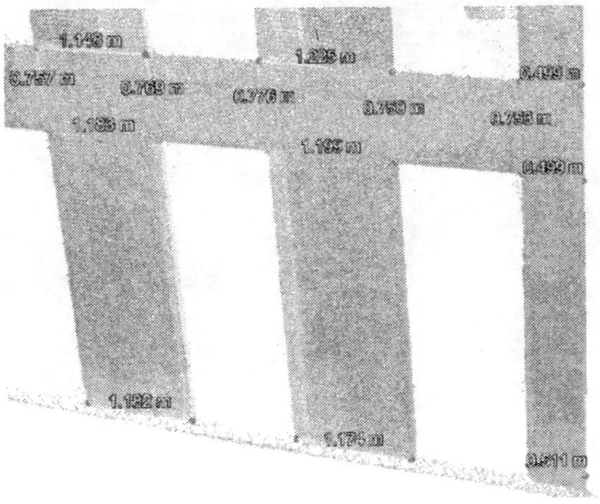

Fig. 9. Examples of windows gaps dimensions.

More challenging was estimate the angular deviations of walls from a reference. Besides, the measurement accuracy was up to 1 cm, the plane matching error has to be estimated. It was computed as standard deviation values for points distances which was taken into consideration and belonging to that plane. With that process, the standard deviation value equals approximately 5 mm. Angular deviations for each wall are presented in Fig. 10 for North, Fig. 11 for South, Fig. 12 for West and Fig. 13 for East. Values are presented in meters.

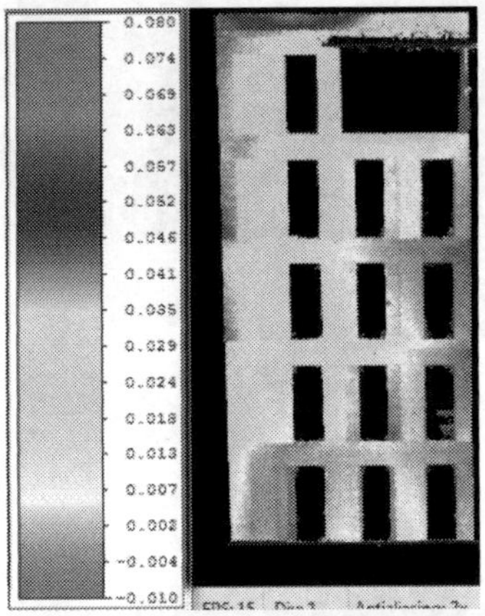

Fig. 10. Angular deviations for North wall in meters.

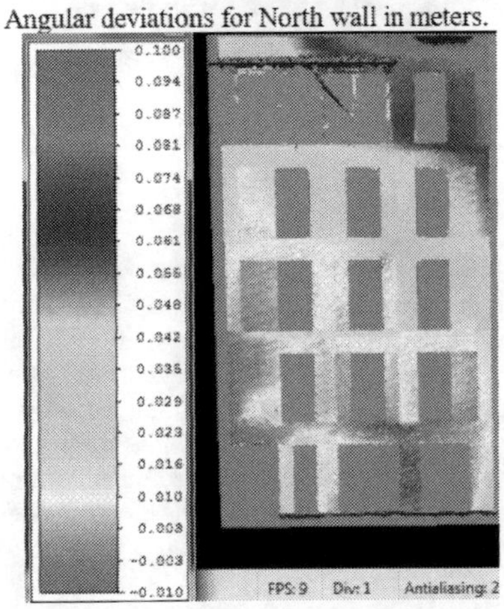

Fig. 11. Angular deviations for South wall in meters.

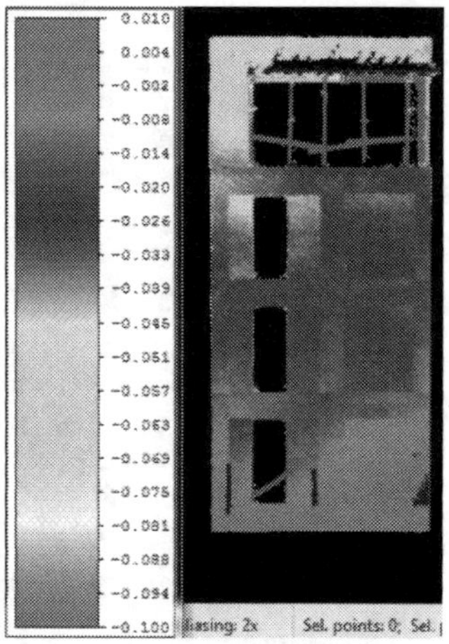

Fig. 12. Angular deviations for West wall in meters.

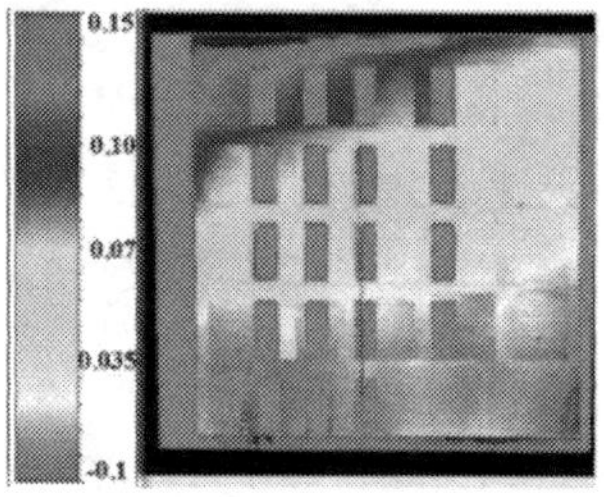

Fig. 13. Angular deviations for East wall in meters.

5 Conclusions

Terrestrial Laser Scanning is an effective tool for as-built inventory performance. It could be used not only in size of a construction estimation but also for digital model creation. Based on that model an angular deviation of the walls could be estimated with high precision and accuracy and based on that data, irregularity could be noticed. Most of the projects are difficult in processes of data acquisition and post- processing but the results could be fully operational. Besides, the laser scanning technology could provide the non- invasion measuring process from could collect over thousands of points per second. It could generate the high accuracy model from the data and low the costs of the data processing at the same time because of the speed of measurements. In comparison to traditional methods where some weight on a rope was hanging from the top of the building, terrestrial laser scanning is an innovative technology where we could collect the whole spatial information about the object of interest with high accuracy. At last, it has to be remembered, that every of new construction has to be evaluate with the real estate markets, especially in urban areas in case of study, described in this article [7][8].

References

1. P. Tysiac, A. Wojtowicz, K. Bobkowska, *INTCESS 2017*, 743 (2017)
2. K. Bobkowska, M. Przyborski, P. Tysiąc, *INTCESS 2017*, 813 (2017)
3. A. Janowski, K. Nagrodzka-Godycka, J. Szulwic, P. Ziolkowski, Computers and Concrete **18 (3)**, 405 (2016)
4. C. Suchocki, J. Katzer, Construction and building materials **122**, 400 (2016)
5. F. Bosche, M. Ahmed, Y. Turkan, C.T. Haas, R. Haas, Automation in Construction **49**, 201 (2015)
6. E. Urbańska-Galewska, K. Żółtowski, J. Ziółko, D. Kowalski, A. Perliński, T. Białek, K. Dobiszewski, M. Miśkiewicz, M. Nowicki, *XXIV Konferencja Naukowo-Techniczna. Awarie Budowlane 2009. Zapobieganie, diagnostyka, naprawy, rekonstrukcje*, 713 (2009) (in Polish)
7. M. Renigier-Biłozor, A. Biłozor A., R. Wiśniewski, Land Use Policy **61**, 511 (2017)
8. M. Renigier-Biłozor, R. Wiśniewski, A. Biłozor, A. Kaklauskas, International Journal of Strategic Property Management **18 (2)**, 198 (2014)

https://doi.org/10.1051/e3sconf/20182600012

The analysis of the accuracy of spatial models using photogrammetric software: Agisoft Photoscan and Pix4D

Adrianna Barbasiewicz[1], *Tadeusz* Widerski[1,*], and *Karol* Daliga[1]

[1]Gdansk University of Technology, Faculty of Civil and Environmental Engineering, Gdansk, Poland

Abstract. This article was created as a result of research conducted within the master thesis. The purpose of the measurements was to analyze the accuracy of the positioning of points by computer programs. Selected software was a specialized computer software dedicated to photogrammetric work. For comparative purposes it was decided to use tools with similar functionality. As the basic parameters that affect the results selected the resolution of the photos on which the key points were searched. In order to determine the location of the determined points, it was decided to follow the photogrammetric resection rule. In order to automate the measurement, the measurement session planning was omitted. The coordinates of the points collected by the tachymetric measure were used as a reference system. The resulting deviations and linear displacements oscillate in millimeters. The visual aspects of the cloud points have also been briefly analyzed.

1 Introduction

Photogrammetric measurements are increasingly becoming an alternative to traditional geodetic measurements. This is because they are faster and easier to perform, and the results are received almost immediately. There are many photogrammetric software, but it is hard to tell from the user panel which of them have more accurate measurement results. Estimation of possible uncertainties generated by software is very important for further analysis of measurement results. Also knowledge of distribution of possible errors is important because we know if we can freely use mean values or not.

In this article were decided to compare two popular photogrammetric software: Agisoft Photoscan and Pix4D.

2 The collection of measurement data

For collecting photos were used the Canon 5D digital camera. Together 126 images were obtained.

3 Used software

Agisoft Photoscan and Pix4D were used for the photogrammetric designation of points. Both programs use still images of stationary objects to generate point clouds, 3D models and orthophotomaps. They are widely used in aviation and engineering photogrammetry, computer animations and remote sensing. With the ability to export processed data to a variety of formats, they are compatible with graphical programs (such as Autocad) or GIS analysis software (such as ArcGis). These programs also have similar hardware and functionality requirements.

4 Development of measurement data

Determining coordinates of the points defined in both programs was similar. First, the pictures were imported into the programs and then merged. Pix4D uses the Initial Processing tool and Agisoft Photoscan - Align Photos. In both cases, it was possible to set the image size to which the similarity points (characteristic, base) were searched. Three levels of accuracy were selected:
• low - 1/4 resolution is taken into account,
• medium - the resolution of 1/2 is taken into account,
• high - a full resolution photo is taken into account.

These functions pre-determine the parameters of the internal and external orientation of the images. As a result, clouds of points without spatial reference were obtained. In order to give georeference, the coordinates of three points of photogrammetric matrix were imported. Minimum three points with known coordinates are required by photogrammetric retroreflection rule. In both programs, exactly the same points were selected and marked exactly on the same location on images. Using Create Markers (Agisoft) and GCP (Ground Control Points) / MTP (Manual Tie Points) Manager (Pix4D), 21 marked points were created and marked on the photos in their actual location. The final stage of the point determination procedure was to optimize the internal and external orientation parameters of the images. In this way coordinates of the point were obtained.

* Corresponding author: tadeusz.widerski@pg.edu.pl

5 Analysis of results

The analysis concerned:
- analysis of calibration parameters,
- determination and analysis of angular and linear displacements,
- visual analysis of obtained images.

5.1 Analysis of calibration parameters

The calibration parameters were different from one another depending on the program used to generate them. There were differences in the location of the main point. The reason for this was the different origin of pixel frame. Agisoft Photoscan determines the location of the points (designated and main) based on the origin of the coordinate system located in the upper left corner of the image. Pix4D takes the lower left corner of the photo as the starting point. Therefore, the focal lengths are different. The biggest difference between the focal lengths, of the order of 4.2386 pixels, was set for the image combining with medium accuracy. It has also been noted that the Agisoft Photoscan program determines four coefficients of radial and tangential distortion, while the Pix4D program determines three coefficients of radial distortion and two tangential distances. In addition, Agisoft's program determines two non-orthogonal affine coefficients. In general, the calculations in both programs are similar. However, there are three major differences between them:
- Determination of distortion corrections using different developments of the Brown model,
- Agisoft Photoscan, when calculating the focal point, takes into account affine deformations,
- Pix4D has already calculated the coordinates of the main point position while Agisoft Photoscan uses a focal offset and adds half the width and length of the image each time.

5.2 Determination and analysis of angular and linear displacements

The next step in the analysis was to determine the values of linear and angular displacements between the coordinates of points determined during the tachymetric measurement and the coordinates determined by the photogrammetric method. During the calculation, it was noted that the Pix4D program did not assign a position of the 12 points determined at the lowest accuracy settings. The reason of that situation was lack of external orientation for 37 images. Therefore, it was decided to skip a comparative analysis of the determined point coordinates obtained from lowest accuracy parameters. Additionally, it was noted that the Pix4D software treats photopoints as fixed points and location values were not estimated (as opposed to Agisoft Photoscan). For linear and angular displacement determined at medium and high accuracy levels, histograms were made and their indices were calculated.

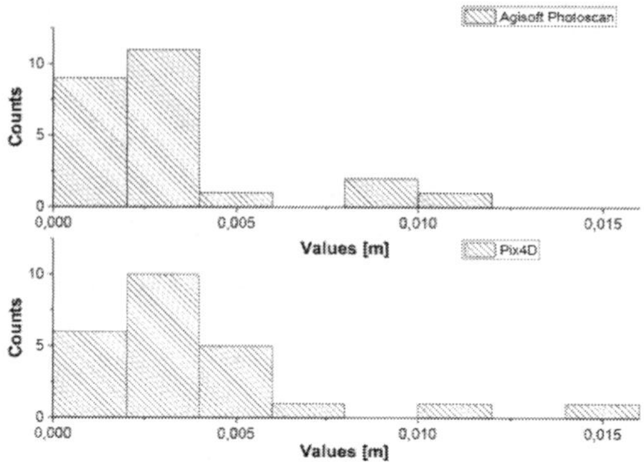

Fig. 1. Linear displacement. Accuracy: medium.

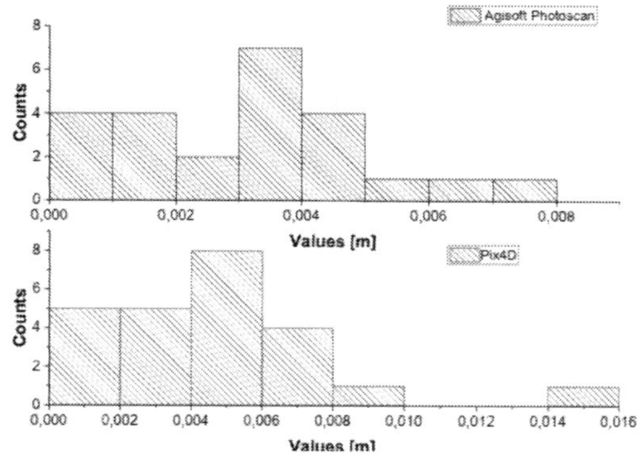

Fig. 2. Linear displacement. Accuracy: high.

Figs. 1 and 2 show histograms of linear displacement values for medium and high accuracy. By comparing the histograms, it can be seen that only histogram made for Agisoft's medium accuracy has a normal distribution. For this reason, the range of +/- 2 σ were used to analyze the coordinate difference vectors, which is around 95 % of the observed values. For this range, the accuracy of the tachimeter is also given. Linear displacements for the remaining accuracy have a right-skewed distribution. This means that the mean and standard deviation are not reliable indicators used in the analysis. Instead, it was decided to use the median and quanta values. In each of the analyzed cases, the 95 % interval is quantified. In Table 1, median values and quantiles for the 95 % stretch are included.

Table 1. Indicators of linear displacement.

	Agisoft Photoscan		
	LOW	MEDIUM	HIGH
R_s [m]	0.00551	0.00321	0.00314
R_{med} [m]	0.00461	0.00337	0.00252
P_1 [m]	-	-0.00069	-
σ [m]	0.00575	0.00195	0.00272
P_2[m]	-	0.00711	-
$Q_{2.5}$ [m]	0.00052	-	0.00038

$Q_{97,5}$ [m]	0.02797	-	0.01017
	Pix4D		
R_s[m]	0.00666	0.00448	0.00375
R_{med}[m]	0.00508	0.00434	0.00287
P_1[m]	-	-	-
σ [m]	0.00793	0.00319	0.00342
P_2[m]	-	-	-
$Q_{2,5}$[m]	0.00000	0.00000	0.00000
$Q_{97,5}$[m]	0.02452	0.01415	0.01429

where:
R_s - average value of offset vector length,
R_{med} - median length of offset vector,
σ - standard deviation,
P_1, P_2 - limit values for 95 % range for medium accuracy (Agisoft),
$Q_{2.5}$, $Q_{97.5}$ - lower and upper interquantile spacing values for the 95 % range.

It was noted that for normal distribution the median value is slightly higher than the mean. In other cases, this dependency is reversed. This is confirmed on the right-skewed side of the slope for the remaining accuracy. It is worth to mention that the right-skewed distribution is not the wrong result. It informs that most of observed values fall below their mean values. For each considered accuracy with right-skewed distribution, all observed values fall within the 95 % range. Thus the $Q_{2.5}$ and $Q_{97.5}$ quintiles are respectively the minimum and maximum values.

To investigate angular shifts, histograms for the azimuthal angles (Fig. 3, Fig. 4) and the horizontal angles (Fig. 5, Fig. 6) were created. At histograms, unit of horizontal axis was marked as *G* which means, in this situation, grads. Full angle measured in grads is equal 400 grads. Azimuthal angles describes directions of horizontal coordinates changes. This changes were calculated by taking into account measured point coordinates and coordinates estimated by software.

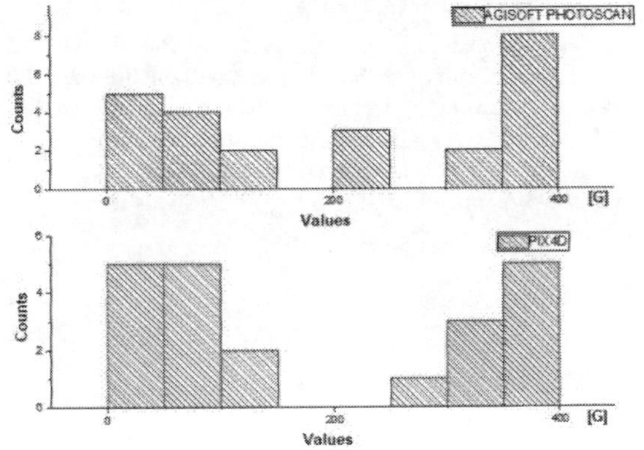

Fig. 3. Histograms of the azimuthal angles. Accuracy: medium.

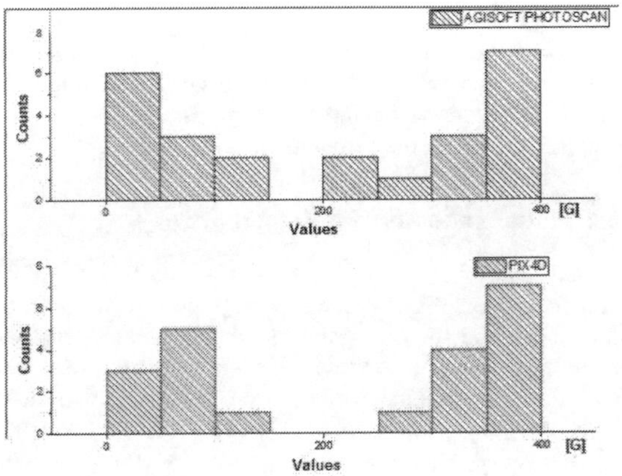

Fig. 4. Histograms of the azimuthal angles. Accuracy: high.

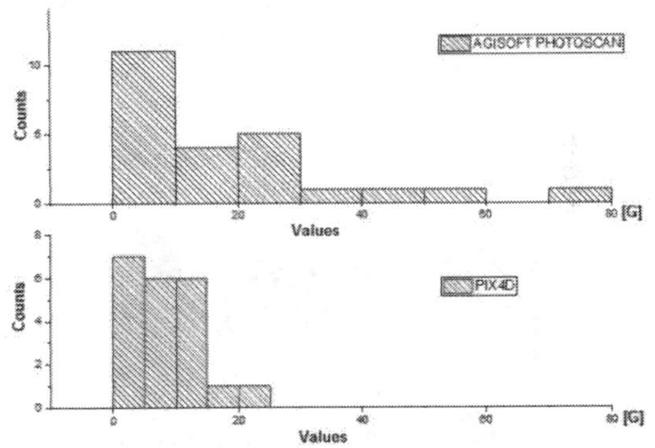

Fig. 5. Histograms of the horizontal angles. Accuracy: medium.

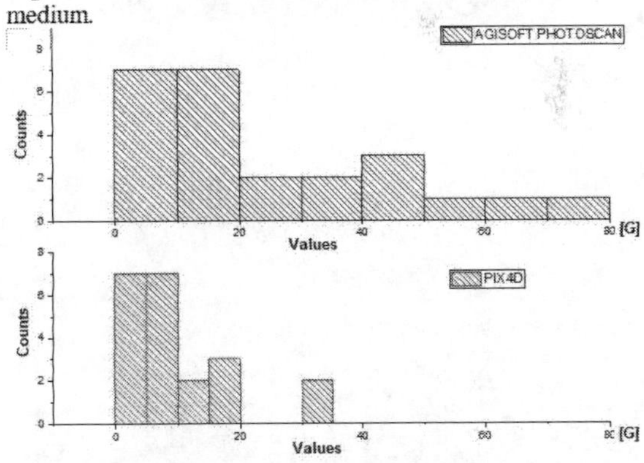

Fig. 6. Histograms of the horizontal angles. Accuracy: high.

On the histograms of the azimuthal angles, bimodal distributions can be observed. The highest incidence rates are between 0 and 100 grads and 300 and 400 grads. This means that the largest azimuthal deviations occur between the north-west and the north-east directions.

The horizontal deviation values for Agisoft Photoscan have a right-skewed distribution. The same accuracy was noted on the histogram for high accuracy in Pix4D. Only the values of the angles determined on the medium accuracy on Pix4D have a normal

 https://doi.org/10.1051/e3sconf/20182600012

distribution. On all histograms, the highest incidence was observed for angles of up to 20 grads. Agisoft Photoscan had set points where the horizontal angular deviation occurs between 0 and 90 grads. For Pix4D, the maximum limit of the compartment was 50 grads.

5.3 Visual analysis of obtained images

Visual analysis focused on the appearance of point clouds and on the number of points generated. It was found that the thickest point cloud was created on the medium accuracy in both computer programs. To confirm this judgment, the exact number of points generated has been verified. On the medium level of accuracy, Agisoft Photoscan created a point cloud composed of 36425 points; on the highest level the point cloud contained 33543 points. Unfortunately, it was not possible to calculate the number of points forming a point cloud in Pix4D, but visual impressions confirm the dependency observed in the first program. In both programs on the high levels of accuracy was a noticeable improvement in the smoothing of points and their concentration. Figs. from 7 to 9 show the appearance of the created point clouds.

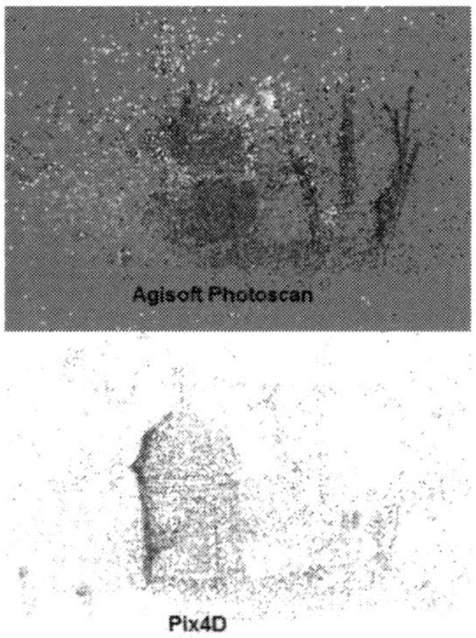

Fig. 7. Point clouds. Accuracy: low.

Fig. 8. Point clouds. Accuracy: medium.

Fig. 9. Point clouds. Accuracy: high.

6 Summary

For both programs received deviations were millimeters. Most of the observed values of angular and linear displacement are within the accuracy range of 95 %. Only the points marked on the images between which a small convergent angle has been maintained, exceeds the set interval.

It has been noted that deviations and angular-linear displacements for medium and high accuracy in both programs are similar. Thanks to this observation, it can be concluded that calculations performed on medium accuracy are also reliable as calculations performed on the high accuracy. Moreover, calculations performed for medium accuracy were faster than performed for high accuracy. It is important when user doesn't have high-end computed and needs good quality of development.

The largest differences in positioning were observed at the lowest accuracy level. Pix4D has not set the position of 12 points determined.

On the basis of the analysis, it was found that satisfactory results can be obtained without the need to mark all photopoints during 3D photogrammetric studies and without exact planning a measurement session. To confirm this thesis additional measurements should be made to prove the reproducibility of the measurement.

https://doi.org/10.1051/e3sconf/20182600012

References

1. Agisoft LLC, *Agisoft PhotoScan User Manual: Professional Edition, Version 1.2* (2016)
2. J. Bernasik, S. Mikrut, *Fotogrametria Inżynieryjna* (Kraków, 2007)
3. M. Korumaz, M. Betti, A. Conti, G. Tucci, G. Bartoli, V. Bonora, A. G. Korumaz, L. Fiorini, Engineering Structures **153**, 224 (2017)
4. Z. Kurczyński, *Fotogrametria* (Wydawnictwo Naukowe PWN SA, Warszawa, 2014)
5. J. McCarthy, Journal of Archaeological Science **43**, 175 (2014)
6. J. A. Morgan, D. J. Brogan, P. A. Nelson, Geomorphology **276**, 125 (2017)
7. Pix4D SA, *Pix4Dmapper 3.2 User Manual* (2017)
8. R. Tokarczyk, I. Stanios, Przegląd geodezyjny **76 (6)**, 7 (2004)
9. Z. Sitek, *Fotogrametria ogólna i inżynieryjna* (PPWK, Warszawa, 1990)
10. A. Wróbel, *Zarys Fotogrametrii* (http://home.agh.edu.pl/~awrobel/resources/Zarys%20fotogrametrii.pdf, access: 2017)

Emotion Recognition - the need for a complete analysis of the phenomenon of expression formation

Katarzyna Bobkowska[1], *Marek* Przyborski[1,*], and *Dariusz* Skorupka[2]

[1]Gdansk University of Technology, Geodesy Department, 80-233 Gdansk, Gabriela Narutowicza St. 11/12, Poland
[2]General Tadeusz Kościuszko Military Academy of Land Forces, Production Department, 51-147 Wroclaw, Czajkowskiego St. 109, Poland

Abstract. This article shows how complex emotions are. This has been proven by the analysis of the changes that occur on the face. The authors present the problem of image analysis for the purpose of identifying emotions. In addition, they point out the importance of recording the phenomenon of the development of emotions on the human face with the use of high-speed cameras, which allows the detection of micro expression. The work that was prepared for this article was based on analyzing the parallax pair correlation coefficients for specific faces. In the article authors proposed to divide the facial image into 8 characteristic segments. With this approach, it was confirmed that at different moments of emotion the pace of expression and the maximum change characteristic of a particular emotion, for each part of the face is different.

1 Introduction

The problem of recognizing emotion based on image analysis is known worldwide. There are many algorithms created by specialists from different fields that allow such analyzes. Examples of such algorithms and analysis are described in the papers [1], [2]. Unfortunately, each of these tools is not reliable. Standard algorithms allow to distinguish only a few basic emotions. It is important to know, there are dozens of emotional states for which there is no algorithm for remotely recognizing these states. To create a reliable tool for remote sensing of emotions, consciousness of the action of the human organism is essential. The big problem is that human facial expressions can be consciously controlled by ourselves [3]. Therefore, the authors consider that observation with high accuracy is important (this translates into high speed of registration) the whole phenomenon of emotion from its beginning (excitement stimulus) to the end. In addition, the analysis of the change occurring on the face should be done without generalizing to one image showing expression. Presented approach will allow you to record all facial changes, including very short-lasting, so-called microexpressions. They are not noticeable by people observing this phenomenon without the use of additional tools, however our brain is able to understand them even without focussing attention. This is caused by the time in which microexpression on the face is visible. Usually there are hundredths of a second. Recent years of research on microexpressions have evolved at a significant pace. It is believed that recognized microexpressions are the key in the process of knowing the true feelings you feel, for example, in people who want to hide such emotions (criminals, terrorists) or in people with mental illness (eg schizophrenia). Research conducted by scientists from various fields shows promising results to detect "hidden emotions" [4-6]. The work of deepening knowledge on this subject can help to create a reliable tool to explore what a person is trying to hide and unconsciously shows on his or her face.

2 Acquisition and processing of data

The authors, have developed an experiment that was based on stimulating emotions (random anger or joy). The stimulus that triggered the emotional state was the randomly selected images from the Karolinska Directed Emotional Faces (KDEF) [7]. 70 people were examined (30 cases were analyzed for this article). A high speed camera was used to register video material. The recording was done at a rate of 1000 frames per second, in grey scale, in high resolution. Each film contained frames representing the face during:

- neutral state,
- stimulation of emotions,
- expression during the emotion
- expulsion of expression.

The resulting videos occupied approximately 6 GB of data on average.

In the study of emotion analysis, high speed cameras are rarely used. Analyzes are made on single images or a stack of images obtained with a standard speed camera. In [8] work, attention has been focused on the need for microexpression analysis with non-standard registration speed. The authors of this work have used rapid

* Corresponding author: marek.przyborski@pg.edu.pl

https://doi.org/10.1051/e3sconf/20182600013

recording cameras at 100 fps and 200 fps. The study data at similar recording speed were collected by the authors of the paper [9]. It can be said that for more accurate measurements and analyzes it is necessary to apply a non-standard registration rate. This has also been confirmed by the authors of this publication in the following articles [10-12].

The initial process of collecting research material and preparing for further analyzes is shown in Fig. 1. Each film was divided into single frames in .tiff format and stored in 8-bit depth as an A_i image - where i stands for the number of frames in the film.

Fig. 2 shows an example of the distribution of a film into individual frames. It was important to normalize the entire research material by stretching the histogram. Therefore it is possible to obtain equally qualitative data. In such a way the prepared images could be subjected to further analyzes.

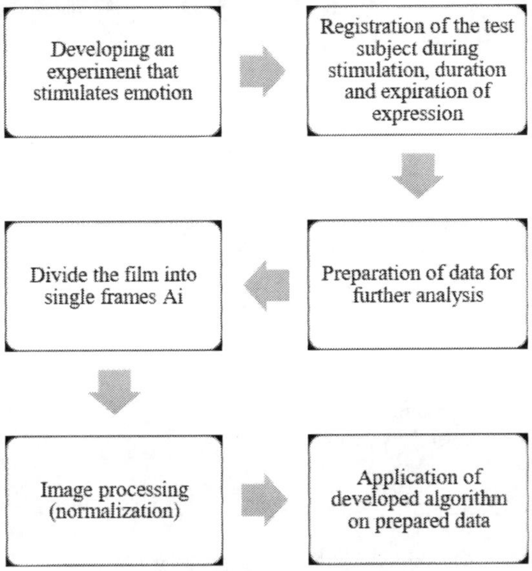

Fig. 1. Registration and preparation of data.

The next step was to develop an algorithm. Fig. 3 shows the flow chart. A completely new approach has been set up. This approach assumes the use of known methods: face detection, mouth detection, eye detection [13], digital image correlation. The initial idea is to divide the face image into specific segments. To achieve this, at the very beginning from the stack of images representing the face with the background (A_i images), the image representing only the face (C_i images) should be cut out. Each of these images for the same movie should have the same size, because the correlation coefficient can only be determined for images of the same size (resolution). Assuming that the image is a matrix, within a single movie, the condition must be fulfilled:

For $C_{1(mn)}$ and $C_{i(gh)}$

$$m = g$$

$$n = h$$

where:
m and g – number of matrix columns,
n and h - number of matrix rows.

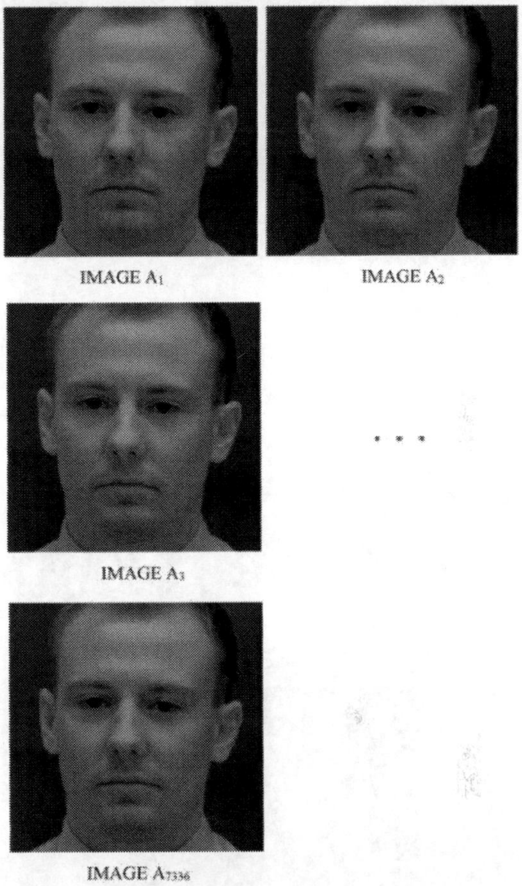

Fig. 2. The effect of unfolding the film on single frames on the example of a film consisting of 7336 frames.

Therefore, the created algorithm assumes the cut, from the original images, the images of a dimension such as the assumed image (A_i) for the first frame. However, the most important aspect was that in every image, the face should be centered. This was achieved by detecting a B_i image representing only the face on all images, while the matrix dimensions of these images were not always the same. Therefore, a dependency has been defined, which allowed to cut images satisfying this condition.

The next stage was eye and mouth detection (Fig. 4 and Fig. 5). This allowed us to create a framework for dividing the frame into eight segments. In this case, the image of the eyes and mouth had to be of the same size as their counterparts for each of the cages. Therefore, for all C_i images, an division was proposed based on the C_1 image identical to detect these parts of the face using existing available functions of Matlab.

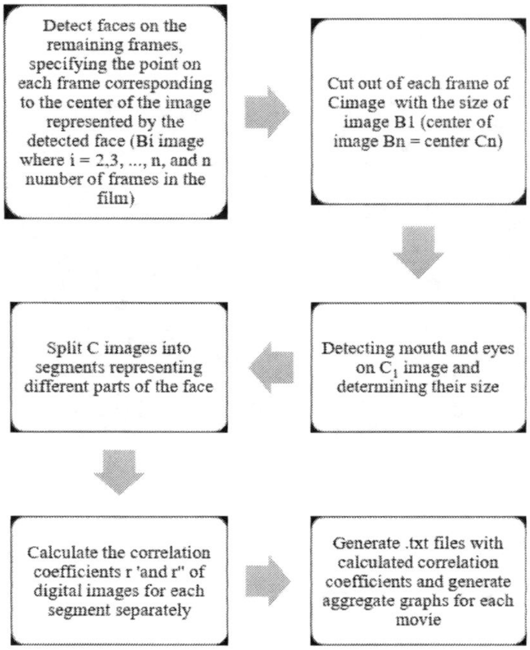

Fig. 3. Description of the procedure used in the algorithm.

Fig. 4. Detected eyes from picture C_1.

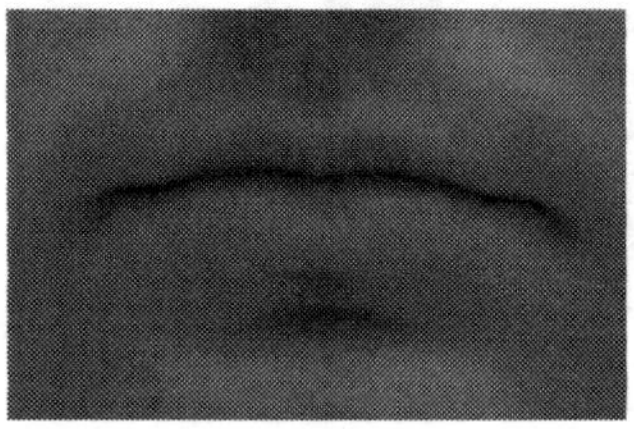

Fig. 5. Detected mouth from picture C_1.

On this basis, an automatic pattern for dividing the facial image into eight segments was created, as shown in Fig. 6. The segments were represented as:

- Segment 1 - lips with a beard
- Segment 2 - lower right cheek,
- Segment 3 - bottom left cheek,
- Segment 4 - eyes,
- Segment 5 - nose,
- Segment 6 - upper right cheek,
- Segment 7 - upper left cheek,
- Segment 8 - forehead.

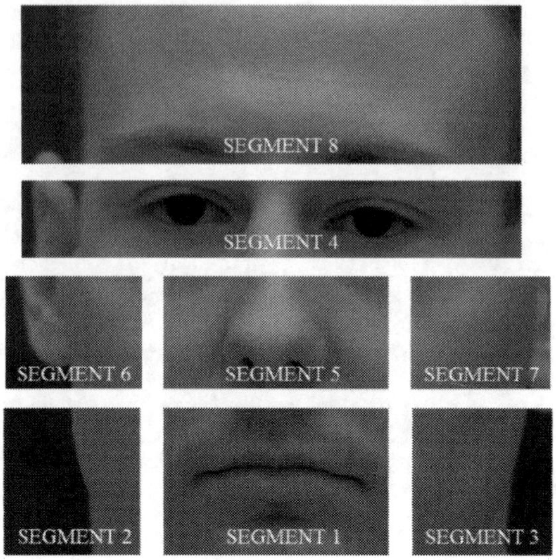

Fig. 6. Scheme of image C1 division.

The next step was to perform the calculation of the correlation coefficient r' and r". The dependencies and formulas have been thoroughly described in other author's publications [10], [11]. The whole process was automatic, using the author's algorithms.
The results generated by the application were used for further analyzes to investigate information about emotion-induced facial changes. Among other things, the speed of the changes taking place on a particular segment was analyzed, the comparison of the attainment of the correlation coefficient for each segment, and the intensity of the change.

3 Results

As a result of calculations have been created matrices of one row containing values of correlation r' (for example $R'_{S1} = [0.994 \quad 0.991 \quad 0.982 \quad \dots \quad 0.976]$, where S1 - segment no 1) and graphs showing changes (Fig. 7) and r'' coefficients for each segment of the test face. The graphs represent a change in the correlation coefficient depending on successive images.

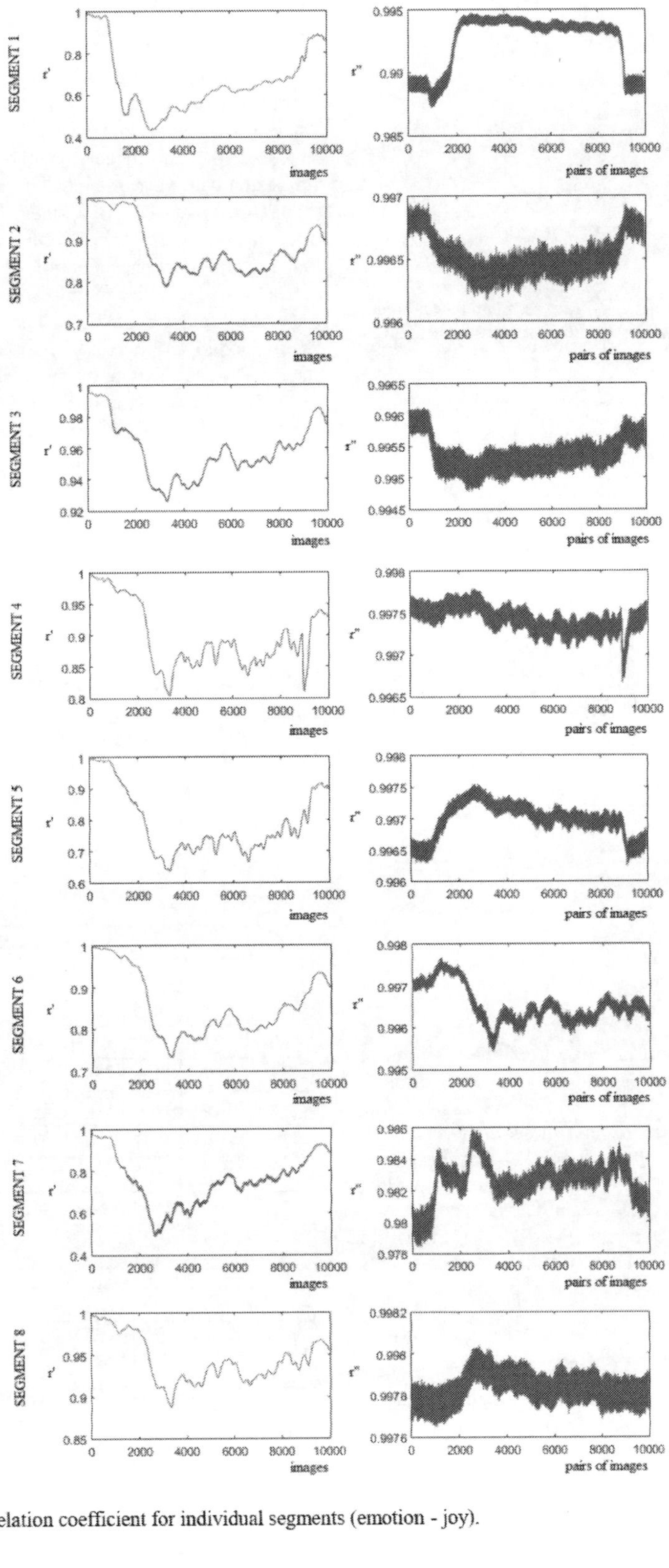

Fig. 7. Variability of correlation coefficient for individual segments (emotion - joy).

 https://doi.org/10.1051/e3sconf/20182600013

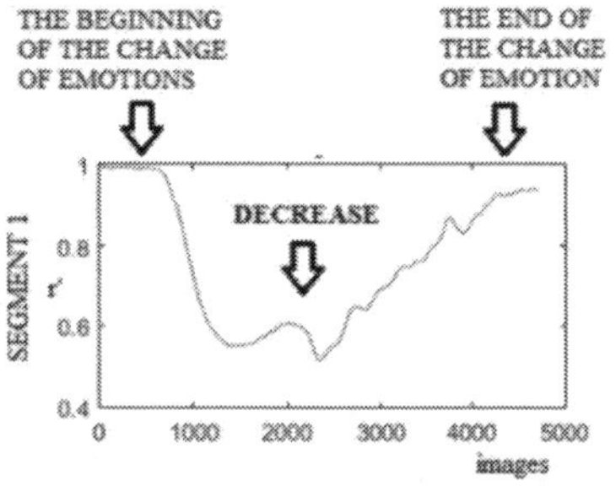

Fig. 8. Interpretation of the graph of the correlation coefficient r' between the first image and sequences (segment 1, emotion - joy).

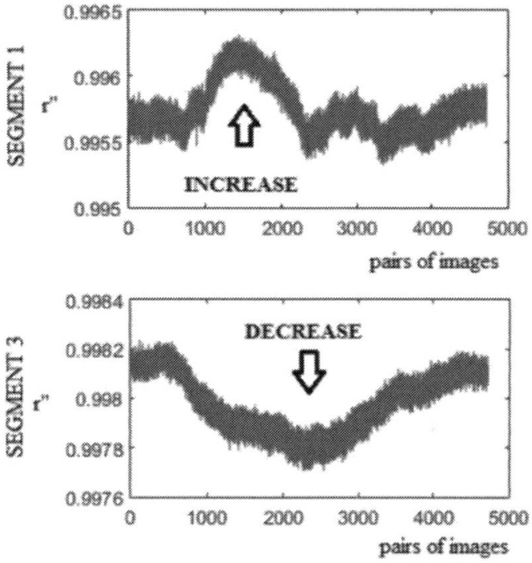

Fig. 9. An example of the increase and decrease of the correlation coefficient between adjacent pairs of images for Segment 1 and Segment 3 (emotion – joy).

Taking into account only the obtained graphics, simple relationships can be noticed. The beginning of change of emotional state and its termination is visualized (Fig. 8), by decreasing the coefficients r' for each segment. The graphs showing the change of the correlation coefficient r'' are different. At the moment of emotion, the drop in coefficient is not always noticeable. In some cases, growth is noted (Fig. 9). The article did not undertake a thorough analysis of data r''. The results are promising and will be used in future research.

3.1 Detection of batting an eye

For all cases, a sudden increase or decrease in the graph showing the correlation coefficient r'' for a given segment represent a rapid change in this area. For example, the files on the graphs generated by the algorithm for segment 4 (representing the eyes of the examined person) show a blink of an eye.

3.2 Comparison of reaching the threshold of decrease in correlation coefficient

By analysing all the graphs for one test person, it is noted that for each segment, the correlation coefficient starts to decrease somewhere else. Based on this, it can be said that not all parts of the face react at the same time and with the same intensity at the stimulus.

Already at the "first glance" the joy of the other person is noticeable, after his smile, which is characteristic of this emotion. And the same is true in case of anger. We recognize it in other people after a specific look. The analyser show that it is not right to focus solely on one of all parts of face. In some cases, before there is a smile on the face representing joy, there are first changes in other areas. To provide this, the following calculations were made:

$$a_k = \overline{R'_{Sk}} \tag{1}$$

$$b_k = \frac{\max(R'_{Sk}) - \min(R'_{Sk})}{2} \tag{2}$$

$$c_k = \max(R'_{Sk}) - (\max(R'_{Sk}) - \min(R'_{Sk}) \cdot 0{,}25) \tag{3}$$

where:
k- segment number
$\overline{R'_{Sk}}$ - mean of the correlation coefficients stored in the matrix R'_{Sk}

In next step, with the command, the number of pairs of images for which values lower than a_k, b_k, c_k were appointed, The results are presented in Tables 1, 2, 3 for example persons. In this way, how quickly responded parts of the face to a given stimulus can be compared.

Table 1. Obtained values a_k.

	Segment no k							
Test	1	2	3	4	5	6	7	8
16	**1236**	2468	2285	2440	2314	2364	1636	2422
28	780	630	645	730	687	**533**	536	712
51	**970**	1196	993	2065	1985	1480	1314	1802
52	1409	1407	1395	781	1484	**586**	1458	1356
22	1644	1232	1407	1303	1228	1211	1315	**940**
24	952	1057	888	1155	885	2651	**775**	2005
59	851	1013	**845**	861	911	915	881	934

Table 2. Obtained values b_k.

	Segment no k							
Test	1	2	3	4	5	6	7	8
16	**1163**	2414	2157	2420	2217	2328	1695	2420
28	809	540	664	793	815	**511**	846	783
51	**974**	2060	1021	2184	2076	2174	2059	2055
52	1396	1407	1405	791	1475	**2**	1416	887
22	4512	1392	1496	1392	1292	1389	1496	**920**
24	868	1077	747	1254	827	961	774	2037
59	**811**	818	775	826	861	841	838	900

Table 3. Obtained values c_k.

	Segment no k							
Test	1	2	3	4	5	6	7	8
16	1003	2227	**964**	2186	1416	2056	1024	2159
28	702	**395**	446	529	646	403	494	513
51	851	1040	**839**	2017	1236	1169	1239	1122
52	1120	1087	1325	585	773	**2**	828	397
22	1286	1112	1162	929	1091	1122	1266	**826**
24	756	728	**427**	1027	754	697	677	1155
59	755	**741**	746	769	801	761	793	815

3.3 Achieving the maximum for the segment

In addition to the fact that at different moments of emotion, the rate of change and the maximum change that presents emotion for each part of the face is different, is selection the following values for the test person - pairs of images representing R'_{Sk}. This dependency for several people is shown in Table 4. The values shown in the table confirm that the maximum facial changes do not occur at the same time for the entire face. Therefore, the analysis of a single photo for the purpose of recognizing emotions is unreasonable because it does not give full information about the expression.

Table 4. Obtained values R'_{Sk}.

	Segment no k							
Test	1	2	3	4	5	6	7	8
16	2673	3268	3268	3324	3337	3318	**2617**	3308
28	961	938	2922	930	944	2922	956	**915**
51	2341	2330	**1385**	2353	2330	2765	2338	2336
52	**1606**	1687	1898	2005	5632	4451	3551	3607
22	5006	1792	1807	1797	1795	1792	1797	**1750**
24	5022	4827	4817	4753	3887	3891	4807	**3877**
59	3772	3896	1085	1359	1104	1474	1105	**1103**

4 Conclusions

The obtained results confirm the assumption that emotional expressions are very complex and require a very detailed and accurate approach to analyses aimed at remote identification of emotions. Given the fact that, not all parts of the face react at the same time and at the same intensity (as has been proven), it is not correct to analyse emotions from individual photos. One image is not a material that can serve as data for a reliable diagnosis of emotion. To get information about the emotional state of a person on the basis of image analysis, it is necessary to observe before the change of emotion, then we have a complete database that allows accurate and thorough analysis of the expression and tracking of changes that take place over time.

It is important to use tools to track changes to get relevant information about the emotions in people with the accuracy of the order of thousandths of a second. In this case, we get the opportunity to catch micro-expressions lasting a few milliseconds that are not controlled by the person being tested. And it is these emotions that reflect the true feelings. The collected material gave another insight into the observations of changes occurring on the face. Analysis for a single movie of several thousand frames, lasting a few seconds, were very time consuming. Therefore, when planning accurate emotional states analysis using high-speed cameras, it is necessary to consider the material handling time. This is of great importance in planning further research related to the analysis of the research material obtained.

References

1. M. F. Valstar, M. Mehu, B. Jiang, M. Pantic, K. Scherer, IEEE Trans. Syst. Man, Cybern. Part B **42(4)**, 966 (2012)
2. D. Datcu, L. J. M. Rothkrantz, *Emot. Recognit. A Pattern Anal. Approach,* chapter 16 (2014)
3. K. N. Ochsner, J. J. Gross, Trends Cogn. Sci., **9(5)**, 242 (2005)
4. T. Pfister, X. Li, G. Zhao, M. Pietik, *Proceedings of 2011 International Conference on Computer Vision,* 1449 (2011)
5. Russell, Tamara A., Chu, Elvina, M. L. Phillips, Br. J. Clin. Psychol. **45(4)**, 579 (2006)
6. Y. J. Liu, J. K. Zhang, W. J. Yan, S. J. Wang, G. Zhao, X. Fu, IEEE Trans. Affect. Comput. **7(4)**, 299 (2016)
7. D. Lundqvist, A. Flykt, A. Öhman, *The Karolinska directed emotional faces (KDEF)* (CD ROM from Department of Clinical Neuroscience, Psychology section, Karolinska Institutet, 1998)
8. X. Li, T. Pfister, X. Huang, G. Zhao, and M. Pietikainen, 2013 10th IEEE International Conference and Workshops on Automatic Face and Gesture Recognition (FG), 1 (2013)
9. Wang, Y., J. See, R. C. W. Phan, Y. H. Oh, PLoS One **10 (5)**, e0124674 (2015)
10. K. Bobkowska, A. Janowski, M. Przyborski, *SGEM2015 Conference Proceedings* **2 (1)**, 1089 (2017)
11. M. Blażek, K. Bobkowska, M. Kaźmierczak, A. Janowski, M. Przyborski, J. Szulwic, *SGEM2017 Conference Proceedings* **17 (21)**, 655 (2017)
12. M. Blazek, J. Janowski, M. Kazmierczak, M. Przyborski, J. Szulwic, *ICERI2014 Proceedings*, 3787 (2014)
13. P. Viola, M. Jones, *Proceedings of the 2001 IEEE Computer Society Conference on Computer Vision and Pattern Recognition* **1**, I-511 (2001)

E3S Web of Conferences **26**, 00014 (2018) https://doi.org/10.1051/e3sconf/20182600014
2017 BGC

The circle object detection with the use of Msplit estimation

Artur Janowski[1,*]

[1] University of Warmia and Mazury, Faculty of Geodesy, Geospatial and Civil Engineering, Institute of Geodesy, Prawochenskiego 15, 10-720 Olsztyn, Poland

Abstract. The paper presents the use of $M_{split(q)}$ - estimation in the filtration and aggregation of point clouds containing a known number of elliptical shapes with preliminary unknown - locations and dimensions. These theoretical solutions may have practical relevance especially in the modelling of terrestrial laser scanning data of objects that have similar shape to circles. Mentioned shapes can be scanned of tree trunks, columns, gutters, etc., that are elliptical in the horizontal plane. The results are satisfied and encourage furthermore detailed research, particularly with the extension of 3D applications.

1 Introduction

Laser scanning measurement technology is popular and well-known technology. This technology is improving and has new applications. It can be specified as:
• SLS (Satellite Laser Scanning) [1],
• TLS (Terrestrial Laser Scanning) [2-4],
• MLS Mobile Laser Scanning [5],
• ALS (Airborne Laser Scanning) [6].

In the case of TLS, the measurement of spatial point locations must be performed from several measurement stations to get complete information, such as the envelope of spatial objects. Despite the efforts, the obtained data set will not fully represent the observed space. It will also contains a lot of surplus data (redundancy) that hinders the assessment of the measured space as well as its numerical analysis.

Therefore, the raw measurement data is usually taken to modelling, which aims to present space as homogeneous sets of qualitatively [7], semantically and functionally objects descripted in the simplest way.
This is usually done by specification the location of the object and its dimensions and orientation in space.

The spatial and topological relations obtained in this way can be used to gain knowledge about the study space as well as to perform analyses based on them such as [8-11].
The modelling process compared to the measurement process is more time consuming and demanding from operator's point of view.

It seems that it can never be fully automated, but many of the tasks can be accelerated enough and performed by the application of a supervised mode.

Nowadays spatial data modelling applications based on TLS measurements include many efficient tools for automatic object recognition. They are based on the nature of the mutual distribution of measurement points, their density and reflectance values [12-16].

However, it is always required to indicate the analysis (modelling) area and the initial parameters determining the operation of the selection, filtering or aggregation tools and the selection of geometric primitives best matched to the modelled element.

The main goal of this article is to present the algorithm for automatic detection, on a given points cloud, assumed number of elliptical shapes (horizontal cross-section).

2 M-estimation

$M_{split(q)}$ [17,18] assumes existence set of n observations l $(l_1, l_2 \ldots l_n)$ that contain q mixed random variables with expected values (1):

$$E\{l_i\}_1 = \boldsymbol{a}_{(1)}\boldsymbol{X}_{(1)},$$

$$E\{l_i\}_2 = \boldsymbol{a}_{(2)}\boldsymbol{X}_{(2)},$$

$$\ldots,$$

$$E\{l_i\}_q = \boldsymbol{a}_{(q)}\boldsymbol{X}_{(q)} \tag{1}$$

where:
$\boldsymbol{a}$ – vectors of known coefficient values,
$\boldsymbol{X}$ – vectors of unknown parameters q random variables.

According to the assumptions of the $M_{split(q)}$, for each observation, the division of potential is defined as the possibility of its belonging to the one of defined functional models. Therefore, each observation li are described by the functional model (2):

$$l_i = E\{l_i\} + v_i = \boldsymbol{a}\boldsymbol{X} + v_i \tag{2}$$

corresponds to q competing functional models (3):

$$l_i = E\{l_i\}_{(q)} + v_{i(q)} = \boldsymbol{a}_{(q)}\boldsymbol{X}_{(q)} + v_{i(q)} \tag{3}$$

* Corresponding author: artur.janowski@geodezja.pl

 https://doi.org/10.1051/e3sconf/20182600014

This assumption allows presents the traditional functional model (2) as a conglomerate of functional models as follow (4):

$$l_i = \boldsymbol{a}_i\boldsymbol{X} + v_i \rightarrow split \rightarrow \begin{cases} l_i = \boldsymbol{a}_{(1)}\boldsymbol{X}_{(1)} + v_{i(1)} \\ l_i = \boldsymbol{a}_{(2)}\boldsymbol{X}_{(2)} + v_{i(2)} \\ \vdots \\ l_i = \boldsymbol{a}_{(q)}\boldsymbol{X}_{(q)} + v_{i(q)} \end{cases} \quad (4)$$

The main assumption of $M_{split(q)}$ is to assign the greatest influence (during estimation process of the particular parameters (5)) those observations that best fit to the particular functional models:

$$\boldsymbol{X}_{(1)}, \boldsymbol{X}_{(2)}, \dots, \boldsymbol{X}_{(q)} \quad (5)$$

and can be used to determine estimators (6):

$$\hat{\boldsymbol{X}}_{(1)}, \hat{\boldsymbol{X}}_{(2)}, \dots, \hat{\boldsymbol{X}}_{(q)} \quad (6)$$

The competition of functional models for every observation adjust them into the whole set of observations according to the conglomerate's assumptions.

Graphically, the idea of $M_{split(q)}$ estimation of the set points, that lays on the plane, of three circles - three functional models ($q = 3$) is shown in Fig. 1.

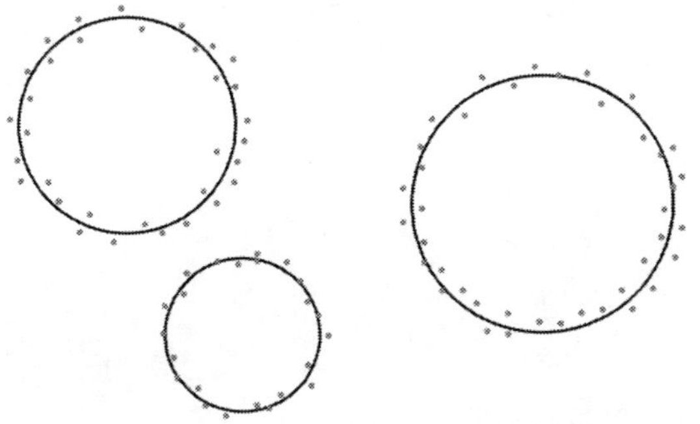

Fig. 1. $M_{split(q)}$ for sets of points belonging to three circles.

The obvious problem in the iterative estimation of the $M_{split(q)}$ is the proper choice of the number of q models in (4). Incorrect assumption of this magnitude will result in the uncertainty of the observation set modelling (the residuals will be too big) and the result will be far from expectation. The number of approximating models can be from 1 to n where n is the number of all observations in the measurement set.

In the presented paper, the choice of the value of the number of models was assumed as a additional task for further considerations. Therefore, in this analysis value of q is the result of the application the statistical method of the cluster analysis with the use of single-neighborhood agglomeration (nearest neighbourhood) [19,20]. This method consists in determining the distance between any clusters as the distance between the two nearest objects (closest neighbours) belonging to distinct clusters. Elements of one cluster form one chain. This method fails with overlapping clusters belonging to different models, then in practical implementations, it may require operator supervision to correct (increase) number of clusters .

After determining the number of clusters and the types of functional models (4) for the all of the $M_{split(q)}$ observations, it can be performed. The result will depend on the correctness of the selected models, their number q, and the relation of the number of observations belonging to each model [21], as well as the way of weighting the observations in the estimation process.

3 Functional model

The use of $M_{split(q)}$ estimation in the elaboration of the 2D or 3D set of points projected to the plane, where they can be treated as q sets of circles requires the use of a suitable functional model. The alignment of a set of 2D points is reduced to determination of the circle centre coordinates and its radius and thus the parameters fulfilling the classical equation:

$$(x_i - x_c)^2 + (y_i - y_c)^2 = R^2 \quad (7)$$

where:
x_C, y_C – centre coordinates of the circle,
x_i, y_i – coordinates of circle points,
R – radius of the circle.

Using the above equation (7) in the LSM (least squares method) estimation requires transformation into a form (8)[22]:

$$P_1x_i + P_2y_i + P_3 = -(x_i^2 + y_i^2) \quad (8)$$

where:

$$P_1 = 2x_c \quad (9)$$

$$P_2 = 2y_c \quad (10)$$

$$P_3 = x_c^2 + y_c^2 - R^2 \quad (11)$$

Such a functional model (8) was chosen for the estimation of q circles in the $M_{split(q)}$ process. Hence:

$$\boldsymbol{L} = \boldsymbol{A}\boldsymbol{X}_{(q)} + \boldsymbol{v}_{(q)} \quad (12)$$

where:

$$\boldsymbol{A} = \begin{bmatrix} x_1 & y_1 & 1 \\ x_2 & y_2 & 1 \\ \vdots & \vdots & 1 \\ x_n & y_n & 1 \end{bmatrix} \quad (13)$$

$$\boldsymbol{X} = \begin{bmatrix} x_1 & y_1 & 1 \\ x_2 & y_2 & 1 \\ \vdots & \vdots & 1 \\ x_n & y_n & 1 \end{bmatrix} \quad (14)$$

$$\boldsymbol{L} = \begin{bmatrix} -(x_1^2 + y_1^2) \\ -(x_2^2 + y_2^2) \\ \vdots \\ -(x_n^2 + y_n^2) \end{bmatrix} \quad (15)$$

$$\boldsymbol{v} = \begin{bmatrix} v_1 \\ v_2 \\ \vdots \\ v_n \end{bmatrix} \quad (16)$$

4 Iterative block of algorithm

An integral part of the algorithm using the $M_{split(q)}$ estimation for segregation of measurement data that represents sets q circles is the "*iterative block of algorithm*" shown in Fig. 2. The input to this block is the data set as the matrices **A**, **L**, **X** (13-15) and parameter q. On such set of data, the LSM estimation is iterated q times. Weights for each estimation are calculated based on the residual values for observations from the previous estimate. The rule here is to assign higher weight values for observations that have higher residual values in the preceding estimation. This rule is implemented by the weight function module (Fig.2).

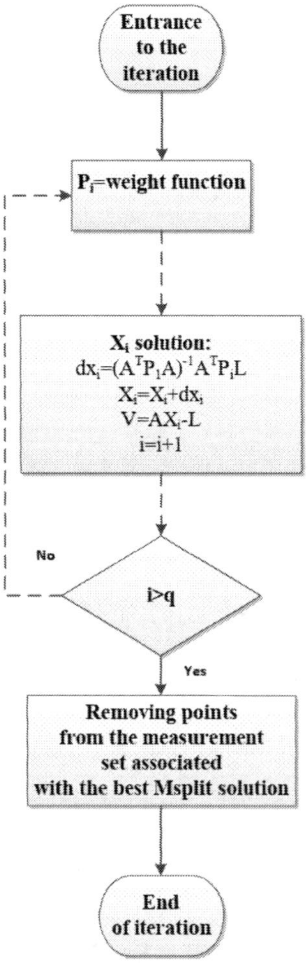

Fig. 2. Iterative block of algorithm.

The method of weighing should be chosen in relation to the optimality of the given task, theoretically and empirically tested. The activity sub-block marked with a dashed line can also be executed iteratively (more than once). Then the weights of the first estimation are obtained by analyzing the residuals of the *qth* estimate of the previous iteration. Such an approach may optimize the execution time of task (based on presented algorithm) for some systems of functional models.

After completing the iterative estimation and obtaining the parameters for the assumed q functional models, it must be chosen the least defective, it means the lowest error fitting into a hypothetical subset of data. A hypothetical subset of data is understood here as a set of observations that Euclidean distance from the found solution of the functional model is less than the assumed value of the parameter τ. The final step of this iterative block is removing, from the measured dataset, observations assigned to the optimal model just found (based on the assumed value of the parameter τ). It means that the remaining observations have a connection with others q-1 solutions.

The iterative block usage process is repeated q times, and each time one optimal solution is selecting and its observations are reducing from all data set. The obtained results of the parameters of all q functional models constitute the approximate parameters of the next iteration that are presented as a dotted line (Fig. 2). The gradient values of these parameters is the basis for the decision to end the algorithm. The schema of main algorithm was depicted on Fig. 3.

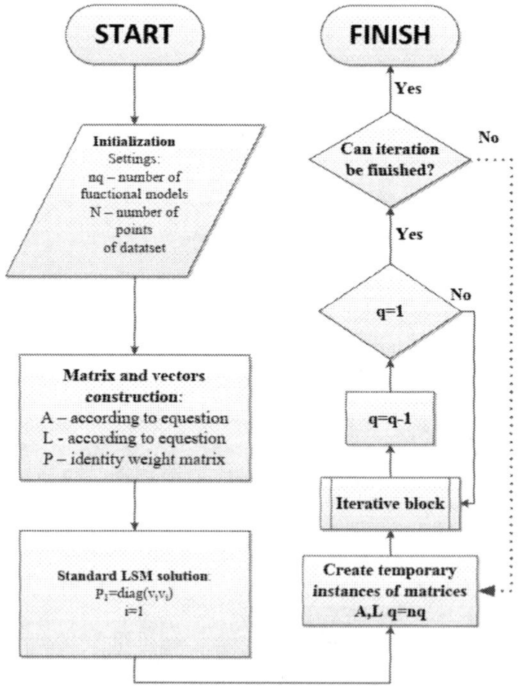

Fig. 3. The schema of main algorithm.

5 Practical implementation

To test the algorithm (Fig. 3), i = 1050 points data was assumed. Points (x_i,y_i) can be interpreted as the creation of 7 functional models in the form of circle equations (8). Three of these circles were overlaid on each, in order to test the reliability of the algorithm to the penetrate results of functional models (Fig. 4).

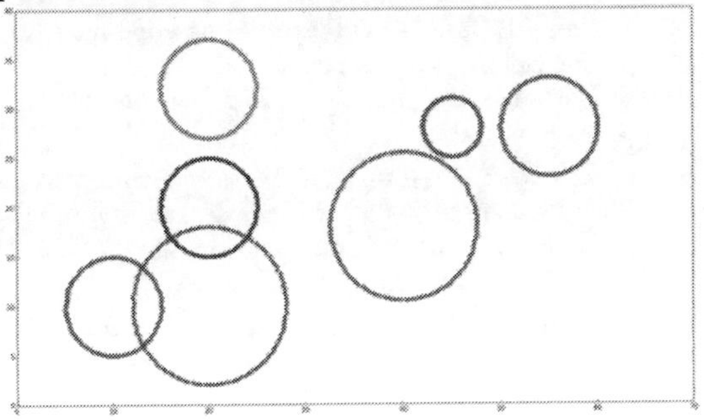

Fig. 4. Visualization of dataset created for validation algorithm purposes.

Parameters of the theoretically defined circles (the theoretical coordinates of the centres of circles x_c, y_c and theoretical radiuses R) with coordinates of points contaminated with Gaussian noise with the maximum value 0.20 m, are presented in Table 1. Coordinates of circle centres and their radiuses, are estimated from data with Gaussian noise, are given respectively as x_{cest}, y_{cest}, R_{est}. Assumed value τ = 1.50 m.

The weighing function was the fiftieth power of residual for observations from the previous estimation.

Table 1. Theoretical and estimated parameters of test circles.

Circle no.	x_c	y_c	R	x_{cest}	y_{cest}	R_{est}
1	10.0	10.0	5.0	10.09	10.00	5.00
2	20.0	10.0	8.0	20.10	9.99	8.00
3	20.0	20.0	5.0	20.11	20.00	5.01
4	20.0	32.0	5.0	20.10	32.00	5.00
5	55.0	28.0	5.0	55.10	28.00	5.00
6	45.0	28.0	3.0	45.10	28.00	3.00
7	40.0	18.0	7.5	40.10	18.0	7.50

As a result of the algorithm use in its 10 main iterations (dotted line in Fig. 3) and increased to 2 iterations in the "iterative block" (dashed line in Fig. 2), one obtains the solution of q = 7 functional models - circles depicted as continuous line in Fig. 5.

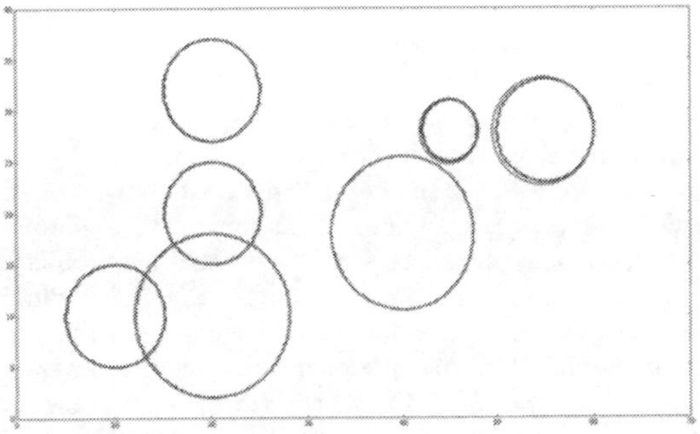

Fig. 5 The solution $M_{split(q)}$ for 7 circles (continuous line).

The obtained values of functional models parameters (x_{mc}, y_{mc}, R_m) and their absolute differences, with the values estimated from the Table 1, are presented in Table 2.

Table 2. Result values for $M_{split(q)}$ with 7 circles.

Circle no.	x_{mc}	y_{mc}	R_m	$\lvert x_{mc}-x_{cest}\rvert$	$\lvert y_{mc}-y_{cest}\rvert$	$\lvert R-R_{est}\rvert$
1	10.10	10.00	5.00	0.01	0.00	0.00
2	20.06	9.99	8.05	0.04	0.00	0.05
3	20.14	19.99	4.99	0.03	0.01	0.02
4	20.07	31.97	5.03	0.03	0.03	0.03
5	54.85	27.90	5.22	0.25	0.10	0.22
6	44.99	28.90	3.13	0.11	0.90	0.13
7	40.09	18.00	7.51	0.01	0.00	0.01

The largest differences from the assumed theoretical values can be seen in solutions 5th and 6th of circle. The numerical justification for such a result is noticed in the order of finding solutions of functional models during using the algorithm. 5th and 6th circles in the iteration process were found earliest then others (there was the greatest negative impact of observation recognized in the further process that belongs to the other solutions).

6 Conclusion

It is worth to notice that among the existing applications of M_{split} [23] there is lack of practical implementation of using this application for more than 2 spited sets (q>2). As the industrial application of the presented algorithm, the measurements of space with a highly structured form, but unknown objects parameters with circle shapes can be indicated. Developing an algorithm for 3D applications would allow the detection of linear infrastructures with circle cross-sections.

Mass acquisition of observation data , which is the domain of most geodetic survey nowadays, forces the search for numerical solutions capable of meeting the expected data processing time and the accuracy of the analysis results.

The proposed modification of algorithm, the extension of the standardized approach of $M_{split(q)}$, works in theoretical considerations. Its practical utility depends on the quality of the data, the number of outliers, the number of observations associated with each of the functional models.

The evident problem of the algorithm needed to solved in the future is the automation of the definition of the value q - the number of functional models. This article assumes its knowing, which greatly influences the versatility of applications. One of potential solution is to place the entire algorithm into another external iteration responsible for the iterative changing value of parameter q. Store of solution results for the accepted value range $q = 1 \ldots maxq$ and choosing this solution where the residual errors will be the smallest. The work involved in the optimal selection of parameter q is the basis for the next study and is closely related to its implementation and commercial applications.

References

1. J. A. B. Rosette, P. R. J. North, J. C. Suárez. Vegetation height estimates for a mixed temperate forest using satellite laser altimetry International Journal of Remote Sensing, 29 (5), 1475 (2008)
2. T. Cao, A. Xiao, L. Wu, L. Mao. Automatic fracture detection based on Terrestrial Laser Scanning data: A new method and case study Computers & Geosciences 106, 209 (2017)
3. P. Tysiąc, A. Wojtowicz, J. Szulwic. Coastal Cliffs Monitoring and Prediction of Displacements Using Terrestial Laser Scanning 2016 Baltic Geodetic Congress (Geomatics), 61, DOI: 10.1109/BGC.Geomatics.2016.20 (2016)
4. K. Bobkowska, A. Janowski, J. Szulwic. 3D modelling of cylindrical shaped objects from lidar data – an assessment based on theoretical modelling and experimental data. Metrology and Measurement Systems, 25(1) (2018)
5. D. Barber, J. Mills, S. Smith-Voysey. Geometric validation of a ground-based mobile laser scanning system ISPRS Journal of Photogrammetry and Remote Sensing 63 (1), 128 (2008)
6. A. Janowski, J. Szulwic, P. Tysiąc, A. Wojtowicz. Airborne and mobile laser scanning in measurements of sea cliffs on the southern Baltic. SGEM2015 Conference Proceedings 2, 17, ISBN 978-619-7105-32-2, ISSN 1314-2704 (2015)
7. M. Walacik. Orthodox valuation method application on the example of coworking SGEM2017 Conference Proceedings 17, 421 (2017)
8. K. Bobkowska, A. Inglot, M. Mikusova, P. Tysiąc. Implementation of spatial information for monitoring and analysis of the area around the port using laser scanning techniques. Polish Maritime Research 24 (s1), 10-15, DOI: 10.1515/pomr-2017-0015 (2017)
9. A. Janowski, J. Rapinski. The Analyzes of PDOP factors for a Zigbee ground – based augmentation systems. Polish Maritime Research, 24(s1), pp. 108-114, DOI: 10.1515/pomr-2017-0028 (2017)
10. M. Renigier-Biłozor. Modern classification system of real estate markets. Geodetski Vestnik. 61(3). DOI: 10.15292/geodetski-vestnik.2017.03.441-460 (2017)
11. M. Bednarczyk. Application possibilities of augmented reality in analog maps Geographic Information Systems Conference and Exhibition "GIS ODYSSEY 2017", 22-27 (2017)
12. C. Wang, P. Hsu. Adaptive building edge detection by combining LiDAR data and aerial images . Journal of Photogrammetry and Remote Sensing 12(4), 365 (2007)
13. A. Janowski, J. Rapiński. M-Split Estimation in Laser Scanning Data Modeling. Journal of the Indian Society of Remote Sensing 42, 15 (2013) DOI: 10.1007/s12524-012-0213-8
14. J. Szulwic, P. Tysiąc. Searching for road deformations using mobile laser scanning. MATEC Web of Conferences 122 (2017) DOI: 10.1051/matecconf/201712204004
15. J. Zheng, J., T. Mccarthy, A. S. Fotheringham,L. Yan. Linear feature extraction of buildings from terrestrial lidar data with morphological techniques. International Archives of the Photogrammetry Remote 37(B), 241 (2008)
16. P. Axelsson. DEM generation from Laser Scanner Data Using Adaptive TIN-Models. The International Archives of Photogrammetry and Remote Sensing 33/B4/1, 110 (2000)
17. Z. Wiśniewski. Estimation of Parameters in a Split Functional Model of Geodetic Observations (Msplit Estimation). Journal of Geodesy 83, 105 (2009)
18. Z. Wiśniewski. Msplit(Q) Estimation: Estimation of Parameters in a Multisplit Functional Model of Geodetic Observations. Journal of Geodesy 84, 355 (2010)
19. M. Renigier-Biłozor, A. Biłozor, R. Wiśniewski. Rating engineering of real estate markets as the condition of urban areas assessment. Land Use Policy 61, 511 (2017)
20. M. Renigier-Biłozor, R. Wiśniewski, A. Biłozor, A. Kaklauskas. Rating methodology for real estate markets – Poland case study. International Journal of Strategic Property Management 18, 198 (2014)
21. A. Biłozor, M. Renigier-Biłozor. Optimization and polyoptimization in the management of land. SGEM2015 Conference Proceedings 2, 1011 (2015)
22. A. Janowski, J. Szulwic, M. Żuk. 3D modelling of liquid fuels base infrastructure for the purpose of visualization and geometrical analysis. SGEM2015 Conference Proceedings 1, 753 (2015)
23. J. Janicka, J. Rapinski. Msplit transformation of coordinates. Survey Review 45(331), 269 (2013)

AUTHOR INDEX